NEW GENERATION WHOLE-LIFE COSTING

NEW GENERATION WHOLE-LIFE COSTING

Property and construction decision-making under uncertainty

Ian Ellingham and William Fawcett

Taylor & Francis
Taylor & Francis Group
LONDON AND NEW YORK

First published 2006 by Taylor & Francis
2 Park Square, Milton Park, Abingdon, Oxon OX14 4RN

Simultaneously published in the USA and Canada
by Taylor & Francis
270 Madison Avenue, New York, NY 10016, USA

Taylor & Francis is an imprint of the Taylor & Francis Group, an informa business

This book is produced from camera-ready copy supplied by the authors.

Printed and bound in Great Britain by Bell & Bain Ltd, Glasgow.

British Library Cataloguing in Publication Data
A catalogue record for this book is available from the British Library

Library of Congress Cataloging-in-Publication Data

Ellingham, Ian.

New generation whole-life costing / Ian Ellingham and William Fawcett.

p. cm.

"The ideas in this book crystallised in the course of two research projects which were supported from 1998 to 2002 by the Partners In Innovation (PII) programme for construction industry research, which was an initiative of the UK Department of Trade and Industry"—Foreword.

Includes bibliographical references and index.

ISBN 0-415-34657-6 (hbk : alk. paper) — ISBN 0-415-34658-4 (pbk : alk. paper) 1. Building—Estimates—Case studies. 2. Life cycle costing—Case studies. 3. Buildings—Great Britain—Maintenance and repair—Cost effectiveness—Case studies. I. Fawcett, William. II. Title.

TH437.E454 2006

692'.5—dc22

2006000853

ISBN10 0-415-34657-6 ISBN13 978-0-415-34657-3 (hbk)
ISBN10 0-415-34658-4 ISBN13 978-0-415-34658-0 (pbk)
ISBN10 0-203-62442-4 ISBN13 978-0-203-62442-5 (ebk)

CONTENTS

Schedule of principles

Schedule of techniques

INTRODUCTION AND ACKNOWLEDGEMENTS

This book is the outcome of two research projects supported by the UK government's Partners In Innovation research programme. An initial study of sustainable refurbishment led to an investigation of new approaches to whole-life costing. The PII projects involved a constructive partnership between government, researchers, property owners and professionals.

Sustainability and whole-life costing

The ideas in this book crystallised in the course of two research projects which were supported from 1998 to 2002 by the Partners In Innovation (PII) programme for construction industry research, which was an initiative of the UK Department of Trade and Industry. Curiously, the questioning of traditional forms of project appraisal emerged in the first project, 'Evaluating sustainable investment in ageing building stock', which was directed at understanding issues of sustainability in the renewal or refurbishment of a mature built environment, as found in many parts of Europe. Many questions were raised about how investments in the building stock are or should be assessed, and the long-term perspective of the sustainability agenda led naturally to the topic of whole-life costing. The low take-up of established techniques of whole-life costing led to speculation about innovative approaches, and these ideas were explored in the second project, 'New generation whole-life costing'.

The PII programme emphasised the dissemination of findings to industry, not the accumulation of academic research reports. This book follows the same practical, applications-oriented approach. There are innovative ideas, but they are put forward as simple and straightforward ways of helping people who are active in the construction industry to ask the right questions about long-term value, and hopefully get useful answers. The very interesting theoretical underpinnings of these new techniques will be addressed in other contexts. Reference is often made to the wisdom embodied in the 'gut feel' of experienced managers. If we help managers or designers – both experienced and inexperienced – make better decisions through enhancing their analytic skills or gut feel, the objectives of the PII project and this book will be achieved.

Organisation of the book

This book has an ambitious scope. It combines an introduction to the basic ideas of whole-life costing (WLC), a critique of the 'standard' method of WLC, and a new approach to WLC. It is aimed at practitioners rather than researchers and most of the material is presented in case format using a modular structure. Each chapter is structured around a story derived from industry partners, presented in a non-technical way. The key concepts are drawn out from the story for clarity. Methods for putting numerical values to the case study examples are described in separate 'technique boxes', which can be skipped on first reading, or by non-mathematical readers.

Chapters 1 to 5 are primarily devoted to basic ideas, including the new idea of lifecycle options. Chapters 6 to 13 are case study applications of lifecycle options. Chapter 14 presents a non-quantified tool for assessing WLC issues in a particular decision context.

So far as possible the chapters are intended to be self-contained so that readers can browse through the book and read sections in any order.

Acknowledgements

As its name suggests, the Partners In Innovation programme involved partnership between government, researchers and practitioners. The project managers for the Department of Trade and Industry were Dr Mervyn Jones and Dr Lawrence Mbugua of Davis Langdon Consulting.

The research component was led by Cambridge Architectural Research Ltd. The CAR team involved Dr William Fawcett, Dr Ian Ellingham, Dr Helen Mulligan and Dr Koen Steemers. Andrew Skeen contributed to the project as a CAR research assistant, together with Fionna Catlin, Dr Jason Palmer and Dr Chris Padbury. The Department of Surveying at Sheffield Hallam University was a research partner in the first PII project, through the contribution of Dr Paul Townsend. He was also involved in formulating the second project, up to the time of his tragically early death. The Department of Land Economy in the University of Cambridge was a research partner in the second PII project, through the contribution of Dr Kanak Patel. This book has relied on input from all members of the research team.

The industry partners covered a wide range of building owners and professionals:

Bennetts Associates, architects: Rab Bennetts

Essex County Council: Chris French and John Parminter

Faithful & Gould, quantity surveyors: Mike Cracknell

John Lewis Partnership: John Carpenter

Newark & Sherwood Energy Agency: David Pickles

Peddars Way Housing Association: Paul Christian

R D Cook Consultants Ltd, architects: Ray Cook

University of Cambridge, Estate Management and Building Service: David Adamson and Martin Dowling.

These were true partners, offering invaluable and probing comment, as well as providing case studies for investigation, while many other individuals assisted the research team's data collection and analysis efforts. These studies led to many of the cases discussed in this book. Accordingly, the cases we present are real, although they have been modified to protect proprietary information, and simplified for clarity.

The following also contributed usefully to the project:

AMEC Capital Projects

Caxton Integrated Services Ltd: Peter Farnan

Diocese of Ely: Dr Matthew Lavis

National Trust (Cambridgeshire): Philip Broadbent-Yale

Stonebridge Housing Action Trust: Anne Byrne.

Finally, numerous seminars were conducted with students, designers and managers in the construction industry, as PII dissemination activities. Feedback from participants in these sessions deepened our understanding of the fascinating range of issues in whole-life costing, and contributed to the content of this book.

The authors

Ian Ellingham BArch MBA MPhil PhD MRAIC trained as an architect at Carleton University in Canada. He studied for the MBA degree at the University of Western Ontario, MPhil at the Department of Land Economy in the University of Cambridge, and PhD at the Department of Architecture in the University of Cambridge. His PhD thesis was 'Consumer influences on the life-cycle of housing'. His work in practice has covered the whole development process in the commercial and voluntary sectors, rather than focusing on the sub-problem of design. He is an associate of Cambridge Architectural Research Ltd.

William Fawcett MA DipArch PhD RIBA trained as an architect at the Department of Architecture in the University of Cambridge, where he went on to PhD research. His PhD thesis was 'A mathematical approach to adaptability in buildings'. As well as working in architectural practice, he has held a lectureship at the University of Hong Kong and was one of the founding directors of Cambridge Architectural Research Ltd. He holds the Chadwick Fellowship in Architecture at Pembroke College, Cambridge.

FOREWORD: FAMILIAR WORDS AND NEW CONCEPTS

The lifecycle option is a key concept in this book's new approach to whole-life costing. It is introduced in Chapter 5. Throughout the book we have restricted the way we use the word option, as described below.

Alternatives and options

In everyday use the words *alternative* and *option* are practically synonymous. When there is a choice, people talk indiscriminately about selecting from the alternatives or from the options; for example, the range of dishes on a menu, destinations for a holiday, or wall finishes for a kitchen might be called alternatives or options. In this book we try to separate the meanings of the two words:

- when making a selection, you decide between *alternatives*
- when holding an *option*, you defer the selection between alternatives to a later date.

Sometimes when making a selection, one of the alternatives may be to retain an option; that is, to defer the selection.

The distinction between alternatives and options is not rigid. Arguably, until the final moment when you come to a decision, all alternatives are options – possibilities for a future decision. In practice it is possible to distinguish between selecting from current alternatives and holding options for future selection.

Option appraisal vs. appraisal of alternatives

Our usage is in direct conflict with what is commonly meant by 'option appraisal' in construction projects. For example, today's issue of an architectural journal refers to a report which 'looks at four separate options' for a new library in Birmingham. We think that the report was looking at four alternatives.

On the other hand, our usage exactly corresponds to the everyday phrase about 'keeping options open'.

We are following a specialised usage of the word option that has a long history. The first English use of the word to describe a contract in which the option-holder has 'the privilege of executing or relinquishing a commercial transaction on terms now fixed' dates back to 1755, according to the *Oxford English Dictionary*. The key point is that the option-holder makes the final decision in the future, rather than having to make it now.

Two words, two meanings

There is an analogy with the words 'robbery' and 'theft' that are practically interchangeable in everyday use: robbery is taking another's property by force or violence, whereas theft is taking another's property by stealth. To most people the distinction may be blurred, as shown by a headline in today's issue of a local paper 'Thieves beat pair in home invasion'. Clearly the malefactors were robbers. Journalists and the public may ignore the distinction between robbery and theft – and why not? – but the police, and the courts, take care to use the words correctly.

In a book about construction decision-making, it seems appropriate to distinguish between alternatives and options.

1

THE RATIONALE FOR PROJECT EVALUATION

Why is it a good idea to carry out a quantified evaluation of construction projects? This chapter explores the reasons, using a conference centre refurbishment project as an example.

A typical construction project

The Chauncey Club is a conference centre located outside the London greenbelt, in a pastoral riverside setting. It is owned by Kingston Oil, a UK-based multi-national company. Since it was acquired in the early 1950s the Club has served well as a setting for training sessions, corporate conferences, team bonding events (when in fashion), and as a place for holidays for frugal employees. It consists of a number of nineteenth century buildings, originally large, elegant houses. One contains the conference rooms, dining areas and administration. The others are used for accommodation, typically containing 20 to 30 rooms.

Recently, the demand for the facility has dropped, and room charges have been reduced to maintain occupancy levels. Managers of the various divisions within Kingston Oil can choose where they hold their conferences and training sessions, and have been electing to book elsewhere. A survey of attendees, together with the centre manager's own observations, revealed that the problem was the poor amenities in the accommodation blocks. Last refurbished in the 1950s, they have shared toilets and bathrooms. Few consumers today want to go down the corridor, towel in hand, to the shower. In particular, because company training sessions are seen partly as a pleasant diversion from the corporate routine, an attractive environment is important. The real cost of conferences is the loss of productive staff time, so the divisional

Refurbishment at the Chauncey Club, where conference delegates are housed in converted nineteenth century villas like this one.

managers are not worried about spending a bit more money on better facilities.

It seems that the Chauncey Club is overdue for refurbishment. Horatio Black, the company's property manager, has been assigned the task of dealing with the project. He has arranged a meeting with Peter Williams, an architect with whom he has worked for many years, and Mrs Van der Vorst, the manager of the conference facility. All three are very experienced and trusted by the company: nevertheless the board of directors expects them to justify any spending proposal for the refurbishment.

Construction decision-making

The Chauncey Club refurbishment is typical of many investment decisions that arise in relation to buildings and property. How do you decide whether to go ahead with a project, and if so, what exactly should you do?

Decision-making for building projects operates in many different ways, and at different levels. Sometimes artistic criteria are dominant, when the investment decision is justified by the aesthetic quality of the finished building. The artist-architect uses some mysterious alchemy to make decisions which (hopefully) add value to the final product. The aesthetic value of a design is fiendishly hard to quantify, but successful examples demonstrate that there is a role for the artist-architect.

Whatever the context, virtually every building project is undertaken in the expectation of future benefits, even though they may be difficult to quantify, as in the case of public monuments like art galleries, opera houses, or olympic stadia. Money spent on such monuments could be spent in other ways. The nature of a society might be revealed by its choice between competing projects: does it build bridges and schools, or palaces and weapons?

Sydney Opera House, designed by Jorn Utzon and completed in 1973, was way over budget and way behind schedule, but those short-term disasters have been more than offset by the long-term prestige of Australia's most iconic building.

In this book we are concentrating on quantifiable aspects of building projects: the resources consumed in construction work, and the measurable costs and benefits that result. Generally we will use money costs, which are a surrogate for every other kind of cost, but the principles would still apply if another yardstick were to be used, perhaps energy. The sums of money are not trivial: even a householder building an extension will be spending tens of thousands of pounds; the Chauncey refurbishment will cost about a million pounds; and today large construction projects cost tens or hundreds of millions of pounds.

Under-investment, over-investment and sustainability

This book is based on the fundamental principle that design and property decision-making should aim for the proper allocation of resources to a construction project. Anything else would constitute under- or over-investment, and implies the waste of resources. Under-investment occurs when a low specification is selected which performs poorly or fails after a short time, compared to a higher specification alternative which would perform better or for longer: if the increased benefits would more than balance the extra investment, then the cheaper, low-performance specification would be inefficient and represent under-investment.

Over-investment is the opposite, when an expensive, high-specification investment is made on the basis of a high level of expected benefits: if it turns out that the benefits are lower than expected and fail to balance the investment, then the expensive, high-specification solution was

Mere cheapness of first cost often enough proves a very expensive delusion. ... With reference to the pecuniary meaning of economy, the young architect must be on his guard. He will be surrounded by those who will press on him materials and methods of construction, which have no merit if they are not cheap. ... Good building can never be what is miscalled cheap building.

E M Barry *Lectures on Architecture Delivered at the Royal Academy* (1881)

Consumption or investment — principle 1.1

Some building projects can be regarded as consumption, the equivalent of buying clothes, holidays or yachts. In these projects it is the customers' privilege to spend just as much or as little as they like. Other building projects, on the other hand, are undertaken as investment, where there is a clear expectation that the resources put into the building will be more than matched by the benefits received.

The distinction is not absolute. Many clients justify putting money into a building project on the basis of both consumption and investment: they enjoy using the building now and also expect to derive benefits in the future.

Building projects are so expensive that they should be very carefully considered, even when a consumer is making the decision between, say, a new kitchen or a holiday in the Caribbean. The consumer's decision will probably be subjective, whereas an investor is more likely to require a quantified project evaluation, typically expressed in financial terms.

inefficient and represents over-investment. Using too many resources in relation to the benefit obtained is as wasteful as not investing enough.

Getting the balance right and avoiding under- or over-investment is a principle that applies to decisions at all scales: a new airport or motorway; a refurbished hospital or school; or a replacement domestic boiler.

It may seem that the authors, as well as most managers, focus on monetary issues and efficiency, and ignore other important aspects of building design, including sustainability. However, the essential nature of sustainability requires the prudent use of resources. The well-known definition of the Brundtland Commission (World Commission on Environment and Development, 1987) states that sustainable development is

Consumption – to repeat the obvious – is the sole end and object of all economic activity. ... We cannot, as a community, provide for future consumption by financial expedients but only by current physical output.

J M Keynes *The General Theory of Employment, Interest and Money* (1936)

Sustainability and efficient investment

principle 1.2

Sustainable development aims to influence today's decisions in order to protect the standard of living and quality of life of future generations. In economic terms this means ensuring an equal or increasing level of consumption by future generations. Their consumption will rely on future production facilities and resources, both natural and man-made. The lesson for the current generation is not necessarily that today's consumption should be reduced, but that we should pass on investments that will produce future benefits.

However, when making investments we use up resources which are irretrievably lost. Sustainable investment should maximise future benefits in relation to the resources used – avoiding resource wastage through under- or over-investment. Under-investment is pure short-sightedness – the failure to invest today in ways that would have produced worthwhile future benefits. Over-investment is just as bad – using up resources for today's investments that fail to deliver future benefits.

It is therefore sustainable to require that resources put into investment should pass a test of efficiency, and that means quantifying inputs and benefits. Different methods of measurement may produce somewhat varying answers, but, if sustainability is a robust criterion, they should converge on the same class of efficient solutions.

In practical applications, inputs and benefits of every description are measured in financial terms. It is often argued that too much weight is given to some factors and too little to others in financially-based appraisal, but the problem is not that costs are measured in money terms, but that incorrect costs are assigned to some of the inputs and benefits. The solution is to assign correct money costs to the resources; there is no necessity to measure in non-money terms. Taxation is important in establishing costs that society believes are fair.

This book describes ways of identifying efficient investment, and avoiding under- or over-investment. This contributes directly to the objective of sustainable development.

'...development that meets the needs of the present without compromising the ability of future generations to meet their own needs.' Clearly, resources have to be consumed in the fulfilment of the needs of today's generation, but they must not be wasted. The goal of sustainability requires efficient investment – that is, the avoidance of under- or over-investment.

Alternative strategies

At the Chauncey Club the need to refurbish the residential accommodation seems self-evident, but the problem is more complex because there are alternative strategies for refurbishment. For example, the improvement might be the minimum necessary to address the worst deficiencies of the present accommodation. Or it might be a more ambitious upgrade to achieve significantly higher quality and more attractive accommodation. Alternatively, the old building could be demolished and replaced with new. And doing nothing, and its consequences, must be included in any comparison.

Peter Williams, the architect, is happy to prepare a design for any of the alternative strategies, together with cost estimates. The more ambitious alternatives naturally require a much higher capital outlay, and Mr Williams would be delighted if this was the chosen approach; but he's not really in a position to know what is best for Kingston Oil. Mrs Van der Vorst, the manager, can estimate the room rentals and occupancy rates for the alternative types of accommodation, with the more ambitious improvements yielding more income; but she's not able to say whether the extra income would justify the higher capital outlay of the ambitious projects. Horatio Black, Kingston Oil's property manager, must consider capital costs, running costs and income levels, and is faced with a multitude of choices. How does he decide what to recommend to the board of directors?

Was this mixture of economy and extravagance an efficient investment? Will Alsop's extension to the Sharp Centre for Design in Toronto, completed in 2004, is a simple box of a building lifted up five stories. It provides functional accommodation and creates an unforgettable image for the Centre.

Experienced managers can often make good project decisions based on the experience of a long career's assimilated information: this is gut feel. Mr Black is approaching retirement and has a hunch about the right strategy, but most of his experience has been built up in

Kingston Oil's commercial and industrial properties, whereas the Chauncey Club is something of a one-off. Also he wonders how his successors at Kingston Oil will get on when they lose his finely-honed intuition. So he thinks this would be a good occasion to carry out some quantified analysis, to confirm his hunch.

The role of quantification and gut feel in decision-making

A number of designers and managers from government and industry co-operated in the research project on project evaluation that led to this book. It became clear from their comments, and from presentations of the research findings, that although quantified evaluation techniques are well known in the property and construction industry they are seldom applied. A variety of reasons were put forward, including major problems in obtaining satisfactory input data, and a recognition that many real estate decision-makers are not fully proficient in mathematical techniques; but, most interestingly, many decision-makers believed that the outcome of analysis often did not accord with their experience and gut feel – they did not have confidence in the results. Hence the search for a 'new generation' of evaluation techniques which designers and managers would use with greater confidence.

Gut feel deserves to be taken seriously – think of it as expert judgement. One stimulus for the research came from two very experienced managers who seemed to be making a curious refurbishment decision. They knew they were going against the recommendations of present-day evaluation methods, but they were sure they were correct although they could not explain why. Using the improved methods of evaluation described in this book we found that they were indeed right. Shortly after participating in our research both managers retired: decades of experience had refined their understanding of building projects, to the extent that they out-performed crude methods of quantification. For the rest of us who want to make good decisions long before retirement, it is highly desirable to turn the insights of gut feel into quantified, repeatable evaluation methods.

Large bridges must be built for generations to come. Engineers today cannot afford to build important structures cheaply, to serve their purpose for the time being, and incur the risk of having to replace them after a short period of usefulness.

Othmar Ammann, bridge designer, writing in 1918 (quoted in D P Billington *The Art of Structural Design: a Swiss Legacy*, 2003). The Sydney Harbour Bridge (above) was opened in 1932.

That is a somewhat idealised picture of gut feel. As with decisions made by artists, it is a black box: you just have to accept the result as it doesn't come with evidence. Different managers will not always agree, but how does one decide whose instincts are reliable?

There must be a role for both quantified methods and gut feel. Quantified methods are useless or worse without insight and judgement, which are the hallmarks of gut feel. And gut feel is not the same as prejudice and complacency: at its best it is built up by constant enquiry and reflection and the consideration of all available information, including quantified modelling. Gut feel and quantified methods should be complementary aspects of top level decision-making.

The first (false) step in quantification

The first candidate method for quantified evaluation is easily identified and easily rejected, although it is often encountered in practice. The method is: choose the alternative with the lowest capital cost.

Descriptive and normative models

principle 1.3

All aspects of human behaviour are complex, and theories of human behaviour are always over-simplified. This certainly applies to economic theory. Simple textbook economics can give very useful insights, but does not provide a complete picture of the world: people tend not to behave in quite the ways that they should do if they followed the simple model of rational economic behaviour.

There are two responses to this mismatch: persuade people to behave differently, or develop more realistic models. The approaches are not really incompatible: if you want to change behaviour, not just describe it, a more realistic model can only help. Think of criminologists – they carry out research to increase understanding of criminal behaviour, in order to find more effective ways of changing it.

Models can thus be classified as descriptive, aiming to make observed behaviour more comprehensible, or normative, aiming to change behaviour for the better.

This book on whole-life costing was in part stimulated by a mismatch between what conventional whole-life costing models say people should do, and what they actually do. We are aiming to build a better model and we hope that people will act on its results, so our ambitions are both descriptive and normative.

Mr Black believes that there are two credible refurbishment strategies for the Chauncey Club: a minimum refurbishment costing £750,000, and a more ambitious project that would cost £1,200,000. Using the 'cheapest-is-best' method the minimum refurbishment project is preferred.

The reason why the cheapest-is-best decision model leads to inefficient investment is that it only takes account of half the picture. There are two sides to an investment decision: capital input in the short term, and benefits derived over a longer period. Both sides are vital. Without considering the benefits, there is absolutely no way of knowing whether a capital input represents under-investment, over-investment or efficient investment.

The cheapest solution may be chosen because there is inadequate capital for a better value solution, or more often because of separation between capital and running cost budgets, when the people responsible for capital want to protect their budget and have nothing to gain from reducing running costs. This kind of management separation, which is all too common, is one of the greatest enemies of whole-life value.

In the special case where alternative solutions under consideration offer exactly the same benefits, the cheapest-is-best method makes sense. This should apply, for example, when evaluating competitive tenders for a fixed project specification. Even here, however, current thinking in the UK construction industry is trying to change price-driven construction procurement, emphasising aspects like quality, reliability and safety. Following current good practice, one must look at a range of benefits and compare them against the price, and not assume that cheapest is the best value.

The rest of this book is devoted to methods of evaluation for construction investment that take account of both costs and benefits. The objective is to identify the best value solutions, not the cheapest solutions.

Low capital cost in a Hong Kong self-built shelter (above) and high capital cost in a City of London office tower (Swiss Re building by Foster & Partners, below). Which was the more efficient investment? It's impossible to know without quantifying both the investment required and the benefits expected.

2

EVALUATING LONG-TERM PROJECTS

The time dimension is critical in evaluating construction investments since buildings are such long-lasting assets. This chapter explains the basic issues related to the time dimension and techniques for handling them. The conference centre refurbishment is used for illustration.

The time dimension: current costs, future benefits

The near-universal situation in construction and real estate projects is that a substantial expenditure is required now or in the short term, in order to receive benefits some time in the future, maybe far in the future. In the case of Kingston Oil's Chauncey Club, the project team is considering expending time and money on a refurbishment within the upcoming few months, in order to achieve the longer term benefit of increased rental rates and level of bookings, and hence greater revenue from the property.

Data is available for both the project costs and benefits. Peter Williams, the architect, can estimate the capital outlay and Mrs Van der Vorst, the manager of the facility, has enough experience to estimate the post-improvement income. The costs and benefits just have to be compared.

Simple payback

It is likely that Horatio Black, Kingston Oil's property manager, will start with a straightforward but imperfect way of comparing costs and benefits – the 'simple payback' method. This involves one easy calculation: just divide the capital outlay by the annual income, to show how many years' of income are required to balance the capital outlay.

Simple payback

principle 2.1

Simple payback is the most basic way of comparing present expenditure with future benefits. It assumes the investment in a building project leads to a regular annual income or saving, for example the rent of an apartment or the fuel savings from insulation. All you do is divide the amount invested by the annual benefit, giving a payback period in years.

If you have investment alternatives you can use the simple payback to compare them. The alternative with the shortest 'payback period' is preferred.

BE WARNED: The only virtue of simple payback is its simplicity. It is a very crude method. The so-called 'payback period' underestimates the time it will actually take to get a return on investment, due to interest charges on capital. Also, simple payback can produce anomalous results. To emphasise this warning we write 'payback period' with inverted commas.

Despite its shortcomings, the simple payback method is sometimes useful for preliminary or approximate analysis. It is not adequate for important investment decisions.

Mr Black has data for two refurbishment alternatives at the Chauncey Club: a minimum refurbishment costing £750,000 which would generate £137,000 income per year; or a more ambitious project that would cost £1,200,000 and generate £170,000 income per year.

Using the simple payback method for the two alternatives, Mr Black quickly establishes that the minimum project has a 'payback period' of 5.5 years, whereas the ambitious project has a 'payback period' of 7 years. So the minimum project is preferred.

Or is it? Mr Black is uneasy. He is aware that the minimum project will not remain attractive for very long, and a further refurbishment would be expected after ten years; whereas the more ambitious project should have a life of fifteen years before the need for upgrading. Simple payback doesn't take account of the project's service life. Mr Black is not yet convinced.

The idea of looking at first cost and running cost together is not new, as shown by this Victorian study of road surfacing in Birmingham:

The following estimates were framed by Mr W Taylor, of the first cost, and the cost for maintenance and cleansing the streets of the greatest traffic in Birmingham, when macadamised, and when paved with granite setts; taking a period of fifteen years for the duration of the granite setts, and taking a vertical wear of 4 inches of macadam annually:

	per square yard
Macadam 6 inches thick, first cost	*1s 6d*
Coating laid twice a year, 4 inches thick annually, 14 years at 1s	*14s 0d*
Cleansing, 15 years at 4d	*5s 0d*
Total cost for macadam, in 15 years	*20s 6d*
Paving, first cost	*7s 6d*
Repair, 15 years	*1s 6d*
Deduct value of old stone	*–2s 6d*
Cleansing, 15 years at 1d	*1s 3d*
Total cost for paving, in 15 years	*7s 9d*

D Kinnear Clark *Recent Practice in the Construction of Roads and Streets* (1890)

Simple payback

technique 2.1

Simple payback compares the capital investment for a construction project (C) with the annual benefit (b), giving the 'payback period' (f). The benefit is typically the project income, which is normally assumed to be the same every year. A short 'payback period' is preferred. The formula is:

$$\text{'payback period'} = \frac{\text{capital cost}}{\text{annual income}} \qquad f = \frac{C}{b}$$

In the Chauncey Club example there are two alternative refurbishment strategies to be compared:

Minimum refurbishment

$C_1 = £750{,}000 \qquad b_1 = £137{,}000/\text{yr} \qquad f_1 = \frac{C_1}{b_1} = 5.47 \text{ yr}$

Ambitious refurbishment

$C_2 = £1{,}200{,}000 \qquad b_2 = £170{,}000/\text{yr} \qquad f_2 = \frac{C_2}{b_2} = 7.01 \text{ yr}$

Using this technique, the minimum refurbishment is preferred because it has a shorter 'payback period'. Simple payback ignores the service life after the end of the 'payback period'.

We put the phrase 'payback period' in inverted commas because it does not actually reflect the financial performance of the investments: if, in this example, the minimum refurbishment generated income for 5.47 years and then the income stopped, the investment would not break even but show a loss. This is because simple payback does not take account of time preference and discounting.

Closely related to simple payback is the idea of *return on investment*, or yield (y). A high yield is preferred. The formula is:

$$\text{return on investment} = \frac{\text{annual income}}{\text{capital cost}} \qquad y = \frac{b}{C}$$

In the Chauncey Club we get the figures:

Minimum refurbishment

$b_1 = £137{,}000/\text{yr} \qquad C_1 = £750{,}000 \qquad y_1 = \frac{b_1}{C_1} = 18.3\%$

Ambitious refurbishment

$b_2 = £170{,}000/\text{yr} \qquad C_2 = £1{,}200{,}000 \qquad y_2 = \frac{b_2}{C_2} = 14.2\%$

The simple payback method and return on investment method will always identify identical 'winning' and 'losing' projects.

Both methods are crude and should only be used for broad-brush comparisons.

Service life

principle 2.2

The service life of a building, or of a component in a building, is the period for which it remains in productive use before being replaced or abandoned. There are a number of reasons why the service life of a building or component reaches an end:

1. physical deterioration: worn out, or more costly to repair than replace
2. functional obsolescence: no longer serves a useful purpose due to activity change or technical progress
3. regulations: incompatible with new regulations
4. loss of esteem: people prefer to change to an alternative
5. external forces: swept away by forces that have nothing to do with the building or component, for example, site cleared for redevelopment.

Estimates of the period of time before condition (1) occurs can be improved by testing the physical performance of buildings or components, taking account of the conditions of use, but exact predictions are impossible.

The other reasons for the end of a building or component's service life depend on wider processes of economic, social and technical change, as well as fashion, and are even harder to predict.

Cashflow and time preference

Mr Black wants an evaluation method for the Chauncey Club refurbishment that takes account of incomes from the refurbished building beyond the end of the 'payback period'. First he needs to set out the project cashflow – a table listing the project costs and the stream of incomes, organised into regular time periods, usually of one year.

Here is the cashflow data for the minimum and ambitious refurbishment projects at the Chauncey Club. Negative numbers are expenditures, positive numbers are incomes. The service life of the minimum refurbishment is expected to be 10 years, so there are ten cashflow entries for income, whereas there are fifteen for the ambitious refurbishment reflecting its longer expected service life.

The current year is usually taken to be time period zero, Year 0. The capital outlay for the refurbishment gives a negative cashflow for Year 0, followed by positive cashflow entries for the years of operation when the refurbished building generates income. This pattern is typical for construction investments.

Life is not long enough – human nature requires quick results, there is a peculiar zest in making money quickly, and remoter gains are discounted by the average man at a very high rate.

J M Keynes *The General Theory of Employment, Interest and Money* (1936)

Having established the project cashflow, how does Mr Black analyse it? It is not very helpful to simply add up all the numbers in the cashflow to get an overall cash balance, because this fails to take account of the time when the transactions occur. Time is extremely important when analysing a cashflow because of *time preference* – the fact that people attach more weight to costs and benefits that occur now or in the near future compared to those that occur further into the future.

Think of a personal analogy: would you rather win a lottery prize this year or the same prize next year? – this year, as you would then be able to enjoy the benefits during the year when you would otherwise be waiting. Similarly for expenditures: would you rather pay a tax bill this year or delay until next year? – next year. Time preference follows from the natural desire to enjoy benefits as soon as possible and to defer disagreeable actions or events for as long as possible.

The established way of deriving the equivalent value today of a future income or expenditure is by *discounting*. The cash value of an income received after one year is reduced by a percentage discount rate to establish its present value. For incomes received after two years the same percentage discount rate is applied twice, and so on for incomes received each succeeding year into the future. The more distant the date when the income will be received, the more times the percentage discount is applied to it.

Year	*Minimum*	*Ambitious*
0	–£750,000	–£1,200,000
1	+£137,000	+£170,000
2	+£137,000	+£170,000
3	+£137,000	+£170,000
4	+£137,000	+£170,000
5	+£137,000	+£170,000
6	+£137,000	+£170,000
7	+£137,000	+£170,000
8	+£137,000	+£170,000
9	+£137,000	+£170,000
10	+£137,000	+£170,000
11		+£170,000
12		+£170,000
13		+£170,000
14		+£170,000
15		+£170,000

Cashflow data for the minimum and ambitious refurbishments at the Chauncey Club.

Cashflow

principle 2.3

The most basic data that is needed for the financial appraisal of a building project is the forecast cashflow. This is simply a list of all the expenditures and incomes, sorted by date of occurrence.

In most building investments there is a large negative cashflow at the beginning, when the building is being built, followed by many smaller positive cashflow entries as the use of the building generates income.

Cashflows are arranged in regular time periods, usually a year at a time for building projects, but longer or shorter time periods could be used where appropriate. All the expenditures and incomes occurring in one time period are aggregated together to give a single number, positive or negative, for that period.

Time preference

principle 2.4

People attach more value to a benefit that is received today compared to the same benefit received a year from now, and much more than if it is received after five, or ten, or more years. Time preference explains why we earn interest on savings, and are willing to pay interest on loans.

Everyone has a time preference, but not everyone has the same time preference. Estimates have been made for the time preference of society as a whole, taking account of the variety of individual views. It has been suggested that there are two components: 'pure' time preference, and the expectation that society will become wealthier over time. This second component is based on the principle that the value of a given benefit, say a £1,000 prize, gets progressively of less consequence to increasingly wealthy recipients. Therefore, a benefit received in the future when society is wealthier will have less value than the same benefit received today. (If people think that society as a whole, or they personally, will be poorer in the future, their time preference will be different. They may value benefits in the future more highly than benefits today.)

Time preference can be regarded as an empirical fact, confirmed by personal experience and numerous experiments, despite the difficulties that arise when one tries to pinpoint its psychological foundations.

It is estimated that in the UK today 'pure' time preference is about 1.5% per year and the rising wealth factor adds another 2% per year, giving society an overall time preference of 3.5% per year.

For a future income with a given cash value, the present value varies according to the size of the percentage discount rate: a high discount rate means that the present value is much lower than the future cash value; with a lower discount rate the present value is closer to the future cash value.

Present value calculations are highly sensitive to the discount rate used. For example, with a 5% per year discount rate, an income with a cash value of £1,000 which will be received in 20 years has a present value of just £377. With a 2% discount rate the present value is £820, and with a 12% discount rate it is £104. For the special case of a zero discount rate, present values and future cash values are equal.

In perspective, objects appear progressively smaller as the distance from the viewer increases. Discounting operates in exactly the same way, with *distance* replaced by *time*: future events make progressively less impact as the time of occurrence moves away from the present.

Just as the effect of perspective in a drawing can be gradual or steep, discounting can also be gradual (low discount rate) or steep (high discount rate).

Discounting

technique 2.2

To get the present value (v_0) of a future income, discounting reduces the cash value (v_i) where the income is received in year i by the discount rate (r) for every time period between now ($t = 0$) and the time when the income or benefit is received. For a single time period the formula is:

$$v_0 = \frac{v_1}{(1 + r)}$$

For more than one time period the discounting is repeated. For two time periods the formula is:

$$v_0 = \frac{v_2}{(1 + r) \times (1 + r)} = \frac{v_2}{(1 + r)^2}$$

The general formula is:

$$v_0 = \frac{v_i}{(1 + r)^i}$$

The impact of discounting is very sensitive to the discount rate used. This can be shown in a table of the present value of £1,000 received at various times in the future. Before personal computers, printed tables like these were widely used in investment appraisal.

	Discount Rate (% per year)						
Year	*0%*	*1%*	*2%*	*5%*	*8%*	*12%*	*20%*
0	£1,000	£1,000	£1,000	£1,000	£1,000	£1,000	£1,000
1	£1,000	£990	£980	£952	£926	£893	£833
2	£1,000	£980	£961	£907	£857	£797	£694
3	£1,000	£971	£942	£864	£794	£712	£579
4	£1,000	£961	£924	£823	£735	£636	£482
5	£1,000	£951	£906	£784	£681	£567	£402
10	£1,000	£905	£820	£614	£463	£322	£162
20	£1,000	£820	£673	£377	£215	£104	£26
50	£1,000	£608	£372	£87	£21	£3	£0
100	£1,000	£370	£138	£8	£0	£0	£0
500	£1,000	£7	£0	£0	£0	£0	£0

Note that the present value of £1,000 in Year 0 is always £1,000 whatever the discount rate; and if there is a zero discount rate the present value of £1,000 is always £1,000 whenever it is received. Otherwise, the present value declines to almost nothing over a long enough time period, even with low discount rates; it declines very rapidly with high discount rates.

Discounting: cash value and present value

principle 2.5

Discounting derives the present value from the cash value of a future income or expenditure by applying a percentage discount rate. The cash value of an income received after one year is reduced by the percentage rate. For incomes received after two years the same percentage discount rate is applied twice, and so on for incomes received each succeeding year into the future. The more distant the date when the income will be received, the more times the percentage discount is applied to it.

Discounting in this way mirrors the effect of interest on bank loans or deposits. If you borrow £1,000 from a bank at 10% interest rate, when the loan is repaid in a year's time you will have to pay £1,000 loan plus £100 interest: the present value of £1,000 is balanced by a future value of £1,100.

Similarly, if you deposit £1,000 savings at the bank at 10% interest, when you withdraw the money in a year's time you receive the £1,000 savings plus £100 interest: again, the present value of £1,000 is balanced by a future value of £1,100.

Discounted cashflow and net present value

Back to Mr Black who now has a cashflow for each of the Chauncey Club refurbishment alternatives. He wants to compare this year's refurbishment expenditure with the present value of the future income streams. To do this, the entries in a cashflow are all discounted to their present values using the project discount rate, and then added up to give the *net present value* (NPV) of the cashflow.

Mr Black will only make an investment if the net present value of the project cashflow, which includes the initial cost, is positive. If the net present value is negative, the future income doesn't justify the investment.

Mr Black is now able to do better than use the simple payback method for the alternative refurbishment cashflows. He first discounts the cashflow at a rate of 6%, and happily both alternatives show a positive net present value: £258,000 for the minimum refurbishment and £451,000 for the ambitious project. They are both viable investments, but the ambitious project has a higher net present value, so it is preferred. This finding is the opposite of the simple payback method, which suggested that the minimum refurbishment performed better. Peter Williams and Mrs Van der Vorst are struggling to keep up, but Mr Black feels he is making progress.

Net present value of a cashflow

technique 2.3

If all the expenditures and incomes in a cashflow (v_0, v_1, v_2, ... v_n) are discounted to their present values using the chosen discount rate (r) and then summed, the result is the net present value of the cashflow.

$$NPV = v_0 + \frac{v_1}{(1 + r)} + \frac{v_2}{(1 + r)^2} + \dots \frac{v_n}{(1 + r)^n}$$

Projects with an NPV above zero are viable. The higher the NPV, the more attractive the project.

For the minimum refurbishment in the Chauncey Club example, the cashflow, discounted cashflow (with a 6% discount rate) and net present value are shown below (table 1):

1. Discounted cashflow for 10 year service life at 6% discount rate

Year	*Cash value*	*Present value*
0	–£750,000	–£750,000
1	£137,000	£129,245
2	£137,000	£121,930
3	£137,000	£115,028
4	£137,000	£108,517
5	£137,000	£102,374
6	£137,000	£96,580
7	£137,000	£91,113
8	£137,000	£85,955
9	£137,000	£81,090
10	£137,000	£76,500
Net Present Value		£258,332

2. Discounted cashflow for 'payback period' at 6% discount rate

Year	*Cash value*	*Present value*
0	–£750,000	–£750,000
1	£137,000	£129,245
2	£137,000	£121,930
3	£137,000	£115,028
4	£137,000	£108,517
5	£137,000	£102,374
6	£68,500	£45,778
Total	£0	–£127,128

The net present value of the ambitious refurbishment can be calculated in the same way, over the project's 15 year service life. With the 6% discount rate the net present value of this cashflow is £451,082. The ambitious refurbishment has a higher NPV and is therefore preferred.

Looking back at the simple payback example, it showed a simple 'payback period' for the minimum refurbishment of 5.5 years. The net present value of 5.5 years of cashflow at 6% discount rate is shown above (table 2 – the sixth year of the cashflow has half of a whole year's income).

The NPV is negative, and it will be negative for any discount rate other than zero: the simple 'payback period' implies a zero discount rate. With higher discount rates, the so-called 'payback period' becomes more and more misleading.

Whole-life costing

The net present value method for evaluating an investment cashflow was first developed about 50 years ago and is widely used in many spheres. In property and construction projects it has acquired a special name: whole-life costing. In fact, whole-life costing is simply the application of net present value analysis to construction projects.

There are some conventions about how whole-life costing for construction is usually done.

First, whole-life costing is normally used to compare alternative specifications for building components, rather than a building project as a whole.

Second, the comparison between alternative components is carried out over a common time period, often 20 to 30 years, regardless of the components' service life. If a component's service life is shorter than the study period, it is assumed that the component is replaced on a like-for-like basis, several times if necessary.

Third, when carrying out a whole-life costing comparison of a component in a building, the cashflow usually records the initial installation costs, the maintenance and running costs, and the replacement costs for the duration of the study period – but no incomes. This is because project incomes can rarely be attributed to individual components. The cashflow is entirely made up of costs: the objective is to find the component with the lowest whole-life cost.

In practice, the purpose of whole-life costing is to identify situations where more expensive components which are longer lasting or have lower running costs are better value than cheaper components, despite the higher initial expenditure. If this did not sometimes occur, then the 'cheapest is best' evaluation rule would be adequate. But although more expensive components are sometimes better value, you cannot assume that this is always true.

There is a risk that advocates of whole-life costing as a method of evaluation can become advocates of high initial expenditure as a general principle. This is wrong – 'expensive is best' is as bad a method of evaluation as 'cheapest is best'. Whole-life costing should be undertaken

Whole-life costing studies usually consider a fixed time period, typically 30 years, rather than extending to the time of a building's demolition.

Consideration of economy in design is often restricted to the examination of ways of reducing first cost, but this is too narrow an approach and it is important also to include the costs of administration, heating, ventilating, lighting and maintaining the building over its life. The equivalent capital value of these may well be as much as, or more than, the first cost of the building itself.

Building Research Station *Principles of Modern Building* (1959)

Whole-life costing

technique 2.4

The conventional use of whole-life costing is for analysing alternative specifications for a building element or component – normally it compares higher specifications which cost more but perform better (B), with lower specifications which are cheaper but do not perform as well (A). The analysis requires cashflow data, which should include any scrap value from disposing of the element or component.

The cashflows can be visualised in a cumulative cost graph. For each specification, the graph starts on the 'now' axis with the first cost and rises gradually with routine maintenance and running costs – the slope is steeper for higher maintenance and running costs. When the component has to be replaced, there is a vertical step corresponding to the replacement cost. For durable components, these steps are more widely spaced.

The graph lines for high-cost, durable components always start off above the lines for cheaper components, but if there are lower costs in use the lines converge, and may cross. After the cross-over point, the higher specification has a lower cumulative cost.

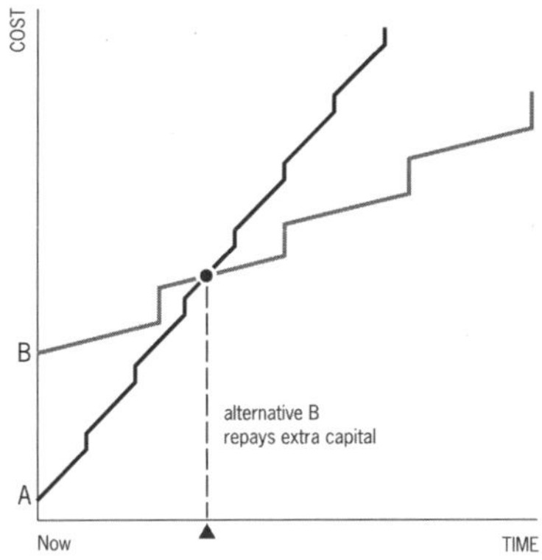

There is one important point to note: the selection between components is not actually made on the cumulative cost of the alternatives, but on net present value which is derived using the appropriate discount rate. The lines on the graph cross because of future running cost savings, and the effect of discounting is to reduce the present value of these savings. Therefore, with discounting the time taken for the lines to cross is longer; and the higher the discount rate, the longer it takes. If the lines do not cross within the time period of the WLC study, the low cost alternative is preferred.

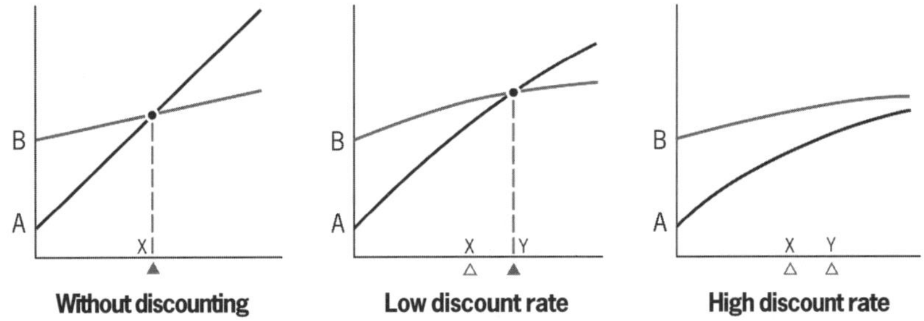

Whole-life costing / Lifecycle costing

principle 2.6

In the construction industry, whole-life costing refers to a technique for assessing both initial cost and future costs of alternative buildings or alternative components, over a study period. A cashflow is established for the study period, and the net present value of the cashflow is calculated with a discount rate. This net present value is called the whole-life cost of the building or building element.

If the service life of an element being studied is shorter than the study period, it is assumed that the element is replaced on a like-for-like basis.

Generally the cashflow consists entirely of costs – installation cost, maintenance and cleaning costs, and replacement cost – so the alternative with the lowest WLC is preferred.

Lifecycle costing is essentially the same technique, except that the cashflow for LCC studies includes production costs before a building component is installed and disposal costs after its service life is finished. The difference between WLC and LCC relates to the cashflow data, not the calculation method.

The purpose of both WLC and LCC is to include longer-term financial implications in the evaluation of design, construction, and real estate decisions. Although the techniques are derived from methods of investment appraisal that are used in other industries, the term 'whole-life costing' is rarely used outside the construction industry.

with an open mind, to discover whether in a specific situation a high initial cost or low initial cost strategy performs best. The result varies from case to case: whole-life costing has not 'failed' when a low initial cost strategy turns out to be preferable.

Importance of the discount rate

Net present value and its application in whole-life costing offer a neat way of taking account of time preference in the evaluation of a project cashflow. The idea of repeatedly applying a discount rate is intuitively plausible and mathematically straightforward. But like all mathematical methods, the answer you get is a reflection of the assumptions you make. The critical assumption for whole-life costing methods is the value of the discount rate.

Horatio Black has an uncomfortable feeling that he chose the 6% discount rate for the Chauncey Club refurbishment rather arbitrarily. With this discount rate he found that the more ambitious refurbishment had a higher net present value than the minimum refurbishment. Suppose the

As long as there is a single hilly railroad track left, at a zero discount rate it would pay to make it level. Why? Because in enough years the savings in fuel would pay for the cost. ... A zero discount rate is like an 'absolute zero temperature' in physics. We can imagine getting close to it, but we can hardly imagine actually reaching the state of a zero rate of discount.

Paul Samuelson *Economics* (7th edn, 1967), with 'discount rate' substituted for 'interest rate' in the original.

Effect of the disount rate

technique 2.5

The following table gives the net present values of the cashflows for the minimum and ambitious refurbishments at the Chauncey Club, for different discount rates:

Discount rate	*Minimum refurbishment*	*Ambitious refurbishment*
0%	£620,000	£1,350,000
2%	£480,614	£984,375
4%	£361,193	£690,126
6%	£258,332	£451,082
8%	£169,281	£255,111
10%	£91,806	£93,034
12%	£24,081	–£42,153
20%	–£175,631	–£405,170
30%	–£326,459	–£644,404

The data can also be drawn on a graph, where the lines representing the NPVs for the two cashflows cross at a discount rate of about 10%.

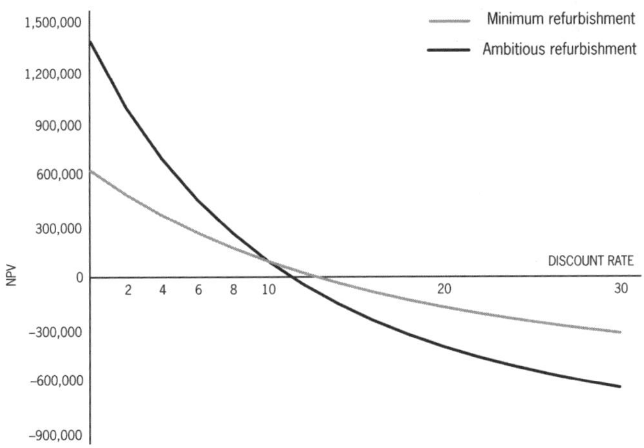

For typical construction projects, NPV falls as the discount rate increases. This is because high discount rates give more weight to the early years in the cashflow – and in construction projects the cashflow starts with a large negative entry, followed by a series of smaller positive entries over the medium and long term.

discount rate should be 10% per year: using this figure, the net present values for the two refurbishment alternatives are virtually identical. If a discount rate of 12% per year is used, the minimum specification performs better than the ambitious alternative. Mr Black finds this disturbing, to put it mildly.

It is very often false economy for anyone (except for the speculative builder who gets rid of badly built houses as soon as they are complete) to stint money on first costs. Too often the money originally saved is spent several times over on patching up an unsatisfactory structure or on maintaining uneconomically finished surfaces. ... And yet there has been very little scientific analysis of the expenditure ... which is economically justifiable and desirable.

J Madge *Tomorrow's Houses* (1946)

The link between discount rate and the outcome of the evaluation is evident. High discount rates give little weight to more distant entries in the cashflow and place great emphasis on the early years, thus favouring a short-term approach and the minimisation of capital investment. Conversely, with low discount rates benefits arising in the future, maybe far into the future if the discount rate is low enough, can justify higher initial costs: low discount rates favour a long-term approach and higher capital investment.

That is the pattern, but what is the right discount rate to use? We discuss this in Chapter 4. In principle, the discount rate is a consequence of the inherent characteristics of the project and the investor: if you fully understand the project and the investor, then you have the necessary information to determine the right discount rate.

3

EVALUATING RISKY PROJECTS

Because it is impossible to predict the future, there is unavoidable uncertainty about long-lasting construction projects. If uncertainty is ignored, the results of project evaluation will be flawed. This chapter explains how uncertainty and project risk can be represented and incorporated into project evaluation. The conference centre refurbishment project continues as the case study for illustration.

Predicting the future

The method of whole-life costing that was described in Chapter 2 involves analysing cashflow data about costs and benefits through the study period, so data is typically required for 20 to 30 years after the construction of a building is finished. Of necessity, this data about future events is estimated – and subject to varying levels of uncertainty. The mathematical calculation of net present value gives very precise answers, but the precision is more or less illusory due to uncertainty in the cashflow data.

In case studies it was found that many managers were dubious about investing to achieve long-term benefits. This should not be surprising. Some enterprises are very transient. Indeed, few high-street shops survive for as long as 10 years without major refurbishment of their interiors, often accompanied by mechanical and electrical alterations. Some substantial buildings are demolished after only a few decades, while conversely, there are many examples of 'temporary' buildings from World War II which remain in service after more than 60 years; and hundreds of thousands of small, cheaply-built Victorian houses are now lovingly conserved after more than 100 years. Needless to say, the original creators could not have predicted their buildings' lives: the buildings were victims,

How long will a building last? A temporary pavilion was built in the City of London in 1951 to help visitors to the Festival of Britain – and it's still there! However, Sir Albert Richardson's design is said to be doomed.

Fifty years is a long time

principle 3.1

Construction decisions are of necessity made within the technical and cost context of their time. The technical and cost context is dynamic, so buildings often survive into very different times. For example, in 1951 the University of Cambridge built a health and sports centre with solid brick walls and single glazing in steel window frames. This made sense when the building was heated by coal costing £2 per ton (shovelled by hand), the insulation then available was expensive and not very efficient 12mm fibreboard, and double-glazing was virtually unheard of. Fifty years later the building is still operating with uninsulated solid brick walls and single glazing, even though fuel prices have escalated and insulating materials and double glazing have become readily available.

The cashflow for a 1951 whole-life costing study would have used 1951 costs for heating the building with coal, but actual events have produced completely different results. The coal boilers were converted to run on oil, and then replaced with gas-fired equipment. No-one in 1951 could have predicted the oil crises of the 1970s with their huge impact on both fuel prices and energy-saving technologies. Is the situation any better for whole-life costing forecasts being made today?

Yet, the health and sports centre has never been upgraded because its life history has been uneventful. The original uses continue virtually unchanged. The building interior feels almost museum-like. It is unusual to encounter buildings that remain essentially unchanged for a half-century.

Major investment is usually stimulated by radical change, frequently due to change of use or the failure of major building elements. There's little pressure to refurbish for 'business as usual'.

or beneficiaries, of evolving circumstances. There is simply a great deal of uncertainty about the future.

Given future uncertainty, Horatio Black, the team leader for the Chauncey Club conference centre refurbishment, is worrying about possible inaccuracies in the cashflow data he used for his first whole-life costing exercise. He is reasonably sure of figures given by Peter Williams, the architect, for the initial outlay on the proposed refurbishment: Mr Black has built many buildings and knows how to interpret estimates and allow for extra costs that always seem to creep in.

However, the post-completion revenues, projected by the conference centre manager Mrs Van der Vorst, might embed some wishful thinking. She recently attended a hospitality industry conference, talked with other conference centre managers and collected their information packs. She found a large range of room rates, occupancy levels and operating costs, reflecting the fact that no two facilities are exactly the same with respect to location,

Cambridge University's health and sports centre dating from 1951 was built to last, and it has. Decisions which were well-suited to 1951 conditions mean that the building is extremely energy-inefficient in a changed world 50 years later.

Retrospective whole-life costing

technique 3.1

The problem with using long-term forecasts is illustrated by the case of Fenners, a university sports and health facility built in 1951. If a whole-life costing study had been done fifty years ago, it would have used 1951 data. For comparison we have added 2001 data and results.

Important differences between 1951 and 2001 include:

- in 1951 the fuel was coal, shovelled by hand, and in 2001 natural gas
- bulidings were heated to lower temperatures in 1951
- few insulation materials were available in 1951, and they were expensive and inefficient compared to the products available in 2001.

The tables show that the 1951 decisions were rational, based on 1951 conditions. But conditions have changed and Fenners is now a greedy energy consumer. The energy crises of the 1970s were unanticipated, but they transformed fuel price levels and consequently led to the development of new fuel-saving components. After two generations, the 1951 building has turned out to be poorly adapted to 2001 conditions.

Is it any easier to make reliable whole-life costing forecasts today than in 1951? Even though we would like to invest in buildings that will be of benefit to future generations, do we know what future generations will want?

Internal design temperatures

	1951	*2001*
Gymnasium	50° F (10° C)	16°C
Office	60-65° F (15.6-18.3° C)	22°C
Schoolroom	55-60° F (12.8-15.6° C)	20°C

External wall insulation

	1951	*2001*
Wall area (m^2)	1126m^2	1126m^2
Insulation	3/4″ fibreboard	30mm polystyrene
Cost (£/m^2)	33p	£7.34
Total cost	£368	£8,273
Value on fuel saved (£/yr)	£7	£1,003
Return on investment	1.90%	12.12%
	Very poor investment	Reasonable investment

Flat roof

	1951	*2001*
Roof area (m^2)	657m^2	657m^2
Insulation	2″ fibreboard	50mm polystyrene
Cost (£/m^2)	£1.03	£10.49
Total cost	£678	£6,893
Value on fuel saved (£/yr)	£33	£4,541
Return on investment	4.87%	65.88%
	Poor investment	Exceptionally good investment

Uncertainty

principle 3.2

Some authors make a distinction between uncertain situations where you have absolutely no idea what will happen, and other situations where you know the range of possible outcomes and have some idea of which are more likely to occur: in the second case you can describe the range of outcomes by a probability distribution.

Most real estate and construction decision making falls into the second category: something is known about the range of possible outcomes and their probabilities.

The probabilities are subjective: they describe the decision maker's state of knowledge or belief about the world, based on the available evidence. When additional evidence arrives, the probabilities are revised.

Strictly speaking, one does not seek probabilities that are a correct description of the world; the goal is to represent accurately a rational belief about the world based on available data. The representation could only be tested against 'real' data, but if 'real' data were available then the rational belief would be changed accordingly.

management, etc. For her estimate she chose numbers somewhat in the middle of the range, based on a comparison between the Chauncey Club and other conference centres. She also assumed that a new fit-out would produce savings in housekeeping.

Mrs Van der Vorst's income estimate is well-researched and credible for current conditions. Forward projections are much more difficult. What will future market trends be? How heavy will the use or abuse be? How will tax, energy, and maintenance costs unfold? How will new technologies influence the building? What about new regulations? How will the setting of the building evolve? Will there be market pressures for change to another use? None of these questions can be answered with certainty.

Mr Black is uncomfortably aware that these and other uncertainties have been ignored in his first whole-life costing exercise, and he wonders whether the results have any value at all. He is eager to find out about tools to deal with these manifold sources of uncertainty.

It is tempting to refine the 'knowable' elements further, for example, getting more accurate pricing for construction costs, but with large uncertainty in other areas this may not improve the quality of the decision to any great extent. It can be observed that project teams often spend considerable effort on refining construction cost estimates,

A Cambridge University teaching building of the 1960s being demolished in 2001, after a very short useful life (by Cambridge University standards). It was out of fashion and expensive to run – like many buildings from the 1960s.

and leave the revenue side at the back-of-an-envelope stage. Perhaps this is because blame is more easily and quickly assigned if a project exceeds its capital budget than if it fails to perform operationally.

Describing and quantifying uncertainty

Almost all forecasts about the future are subject to uncertainty, but the amount of uncertainty varies greatly. Predictions of future lunar eclipses, for example, are as good as certain, relying on the regularity of the solar system. Predictions of heads or tails for a single toss of a coin are utterly pointless because, without cheating, the outcome is random. Forecasts about events in the life of buildings fall between these extremes.

Fan diagram

A neat way of indicating the level of future uncertainty is the fan diagram. The diagram starts with a single point

Fan of uncertainty

principle 3.3

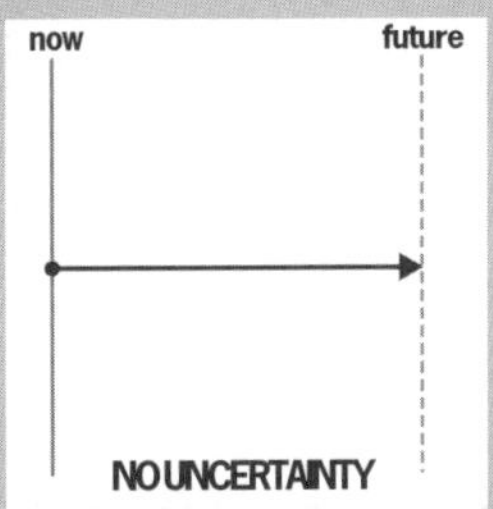

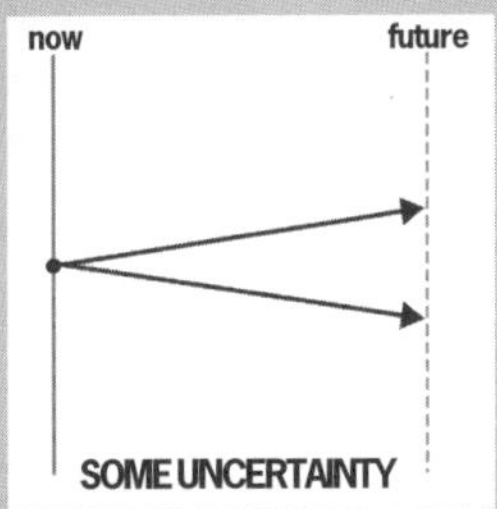

In the 'fan' diagram time runs from left to right – a vertical axis on the left represents 'now'. The up-down dimension describes the state of the system of interest – say, the price of natural gas. We know what is happening today, so today's price is marked with a single point on the 'now' line. The price may go up – the line which rises from today's price; or it may go down – the falling line. In the future, towards the right hand side of the diagram, there is a range of possible values, shown by the difference between the rising line and the falling line.

In situations where the future is predictable (there must be some?) there is just one line from today's value; it may be horizontal if nothing changes, or it may rise or fall in a fully predictable way.

The amount of uncertainty is represented by the divergence between the rising and falling lines – they are close together when there is a small amount of uncertainty and widely divergent when there is a lot of uncertainty.

representing today's known situation. If the future is certain, we draw a single line for future values. If there is a small amount of uncertainty, we draw two gradually diverging lines and if there is a lot of uncertainty the two lines diverge steeply.

Binomial tree

The fan diagram can be developed into a *binomial tree* to give a quantified picture of uncertainty.

To construct a binomial tree for the future income of the Chauncey Club, Mr Black begins with the current value and then assesses how much it could change in, say, one year. Mrs Van der Vorst is confident that an income of £137,000 per year could be achieved when the refurbishment opens. But then the market could change, due,

Binomial tree

principle 3.4

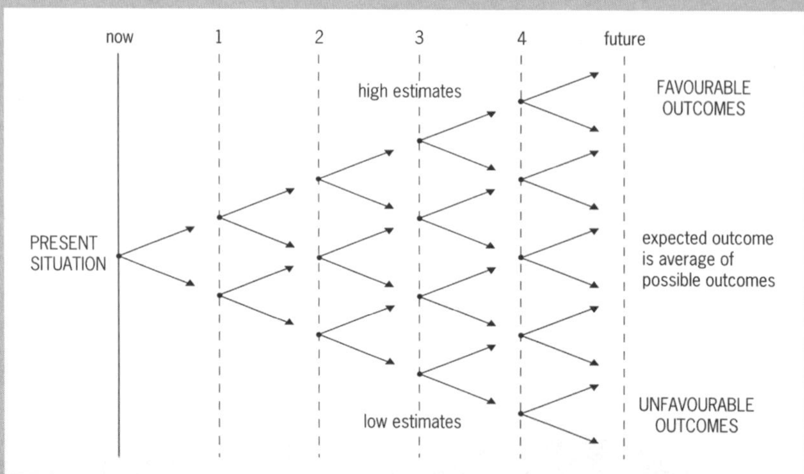

Like the fan diagram, the binomial tree starts with today's known value on the 'now' axis. Time, running to the right, is divided up into equal time periods – usually a year for analysing construction projects. Over each time period, the range of feasible outcomes must be estimated, beginning with today's value. The highest and lowest values after one time period are marked in the next time period. The pattern repeats throughout the binomial tree, with the difference between the highest and lowest feasible values getting progressively wider.

The same ratio of upward and downward movements compared to the current value is normally used throughout the binomial tree.

The binomial tree is a forecast of future uncertainty, not a prediction of future outcomes.

Binomial tree

technique 3.2

The binomial tree is a quantified version of the fan diagram for representing future uncertainty. The time axis of the diagram is divided into regular time periods. Starting with today's known value at the time period 0, an up and a down line are drawn to time period 1; the up line leads to the highest credible value after one time period and the down line leads to the lowest credible value. The probabilities are marked on the two lines – they must add to 1 but they don't have to be 0.5 and 0.5. From each of the two values at time period 1 new pairs of branching lines are drawn to time period 2, in the same way. And so on for the rest of the tree.

The income forecast for the Chauncey Club can be drawn as a binomial tree. The nodes' probabilities of occurrence in each year are marked on the tree:

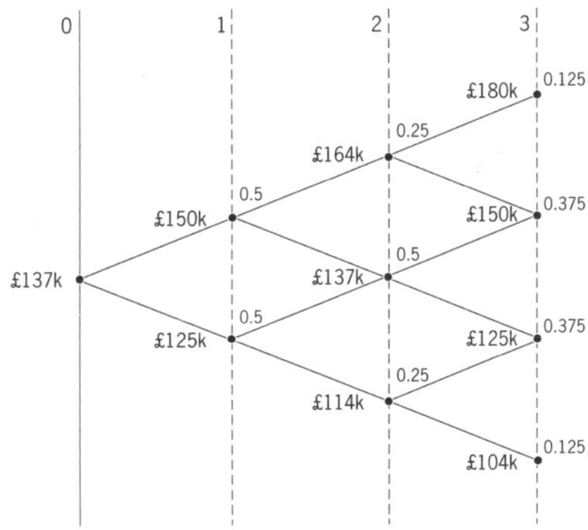

In this example the rate of increase and decrease between time periods is: 9.5% or 0.095. Thus in the first branch:

$$\text{up value} = £137{,}000 \times (1 + 0.095) = £150{,}000$$

$$\text{down value} = \frac{£137{,}000}{(1 + 0.095)} = £125{,}000$$

Because of the regularity in the binomial tree, the middle value in time period 3 can be reached either by a rise between time periods 0 and 1 and a fall between periods 1 and 2, or by a fall between time periods 0 and 1 and a rise between periods 1 and 2. It is called a recombining tree.

The number of possible values increases with each time period, but the central values are more probable than the extreme values. The probabilities attached to each value can be calculated, starting from the left, where the starting value is certain, ie. probability = 1. The probabilities of the values in the second time period are given by the weights attached to the branches in the binomial tree (0.5 and 0.5 for the two values in this case). After that, the probability of a particular value is given by the probabilities of its direct antecedents to the left (either one or two of them) multiplied by the probabilities attached to the connecting branches. These probabilities are marked.

for example, to the economic cycle or to competing conference centres opening (or closing). Mrs Van der Vorst tells Mr Black that in the second year she believes that the annual income could rise to £150,000 in the most favourable circumstances, or drop to £124,000 if trading is difficult. She thinks the two outcomes are equally probable.

Mr Black marks the high and low forecasts in his binomial tree. In following years the income could continue to rise, or fall, or reverse direction, so the range of possible incomes becomes wider. The uncertainty surrounding Mrs Van der Vorst's income forecast consequently increases for each succeeding year.

Expected value

The weighted average of the upward and downward trends in Mr Black's binomial tree is £137,000 – this is the *expected value* of the uncertain outcome: it is the same as Mrs Van der Vorst's original forecast, but there is a big difference between a prediction and an expected value. The binomial tree gives Mr Black a clearer idea of the extent of uncertainty faced by the refurbishment project.

Volatility

How does one estimate the range of possible future values when constructing a binomial tree? One has to rely on past experience, even though it is not an infallible guide.

In 1960 when the architects Chamberlin Powell & Bon designed New Hall, a new women's college in Cambridge (above), they tried to take a long-term view of the choice of fuel for heating. However, their predictions have proved completely wrong (so far):

New Hall is intended to last for many generations. It seems likely that, as time goes by, the cost of electricity will go down. Coal may well become too expensive to use as a fuel, and the supply of cheap oil may well become exhausted; by that time the existence of more efficient nuclear power stations may be anticipated. Although electric heating is more expensive today, this difference is not likely to last indefinitely.

Expected value

principle 3.5

When you can't predict and exact outcome, but you do know the range of possible outcomes and their probabilities, you can work out the expected value: it is the average of all possible outcomes, weighted by their probability.

If you toss a coin and win £10 if it lands 'heads' and nothing if it's 'tails', the expected value is £5 – the average of the two equally probable outcomes, £10 and zero.

The expected value is only an average of many possible outcomes: it is not a prediction of the *actual* outcome. It is just possible that the actual outcome turns out to be the same as the expected value – but normally the actual outcome will be either higher or lower than the expected value.

Expected value technique 3.3

When a precise value is unknown, but the range of possibilities and the probability of each possibility is known, then the expected value can be calculated. It is the average of the possibilities, weighted by their probability.

If there are just two outcomes, and they each have a probability of 0.5, then the expected value is the average of the two outcomes.

If the possible values are:

$x_1, x_2, x_3, \ldots x_n$

with the probabilities

$p_1, p_2, p_3, \ldots p_n$, where $p_1 + p_2 + p_3 + \ldots p_n = 1$

then the expected value is $(x_1 \times p_1) + (x_2 \times p_2) + (x_3 \times p_3) + \ldots (x_n \times p_n)$

For example, taking the data from the Chauncey Club binomial tree at Time 3, there are four possible values:

£180k, £150k, 125k, £104k

with the probabilities:

0.125, 0.375, 0.375, 0.125, which add to 1

then the expected value is:

(180 × 0.125) + (150 × 0.375) + (125 × 0.375) + (104 × 0.125) = 138, or £138k

This is very close to initial value of £137k because the upward and downward branches in the binomial tree are equal.

Many probability distributions are of this type, with values in the middle of the range having a higher probability than the outlying values. Drawn as a graph this gives the familiar bell-shaped curve. The expected value is at or near the highest point on the graph.

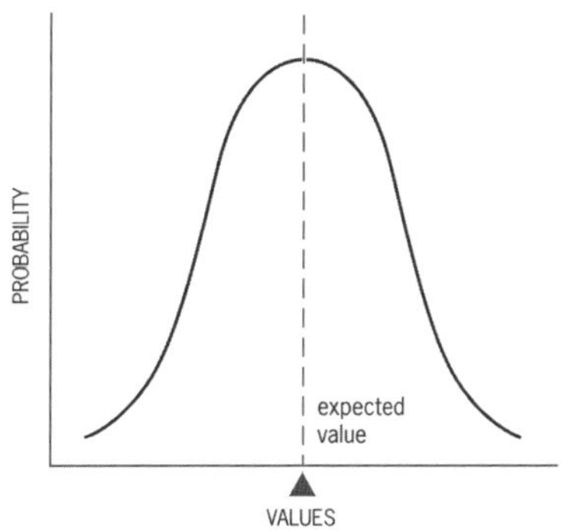

Volatility

principle 3.6

Volatility is a measure of change over time. Phenomena that change often and radically have high volatility, whereas those that change little and slowly have low volatility. Volatility can be measured precisely from historic data. Past volatility is often used as a guide to future volatility. It does not allow the future to be predicted, but it suggests the range of possible future outcomes – the divergence in the fan diagram or binomial tree.

Past experience tells us whether the value in question has been stable with limited change, or unstable with pronounced change. This property is *volatility*. For example, the price of a commodity like natural gas goes up in some years and down in others depending on world politics, competition between suppliers, and the coldness of winters. This past volatility can be measured. The price of gas is likely to go on changing in the future, making next year's price uncertain; but we can use the past volatility to estimate the amount of change that could be expected next year, and therefore the highest and lowest prices that are reasonable for the binomial tree.

Volatility does not just apply to prices. There is volatility in the profitability of oil companies, in conference centre occupancy rates, the number of hours of sunlight per year (important for solar energy), and the number of days of winter frost (important for fuel-saving strategies).

Responses to uncertainty

How does Mr Black's evaluation of the Chauncey Club refurbishment project change when he understands the uncertainty surrounding Mrs Van der Vorst's income projections? Like most people, Mr Black is risk-averse and he finds an uncertain investment with an expected value of £137,000 per year a less attractive proposition than a certain investment with a value of exactly £137,000 per year. The wider the range of uncertainty around the expected value, the greater the shortfall compared to a certain investment which is equal to the expected value.

Volatility

technique 3.4

Volatility is a description of the amount of change over time – the frequency and magnitude of upward and downward movements. Things (people, politics, prices, etc) which are stable have low volatility; things which vary rapidly and dramatically have high volatility. When forecasting, there is more uncertainty about the outcome where there is high volatility, so volatility is a measure of uncertainty.

Looking backwards from the present, it is straightforward to measure volatility from historic data. Past volatility is often used for forecasting. For example, it is reasonable to forecast that the future price of natural gas will move upwards or downwards in the same way as it has in the past.

The volatility of a data series is measured by its standard deviation (s), which is the square roots of its variance (s^2). For a data series x_1, x_2, x_3, ... x_n with a mean value of $\bar{x}$:

$$\text{variance } s^2 = \frac{(x_1 - \bar{x})^2 + (x_2 - \bar{x})^2 + (x_3 - \bar{x})^2 + \ldots (x_n - \bar{x})^2}{(n - 1)}$$

$$\text{volatility } s = \sqrt{\frac{(x_1 - \bar{x})^2 + (x_2 - \bar{x})^2 + (x_3 - \bar{x})^2 + \ldots (x_n - \bar{x})^2}{(n - 1)}}$$

For example, we can calculate the volatility of two data series of seven values, say, the number of rooms booked per night over the last week in two hotels:

Hotel Acropolis:	5, 6, 4, 5, 7, 3, 5
Hotel Bristol:	5, 7, 3, 5, 9, 1, 5

Both have a mean of 5 bookings per night, but the number of bookings in Hotel B swings twice as widely as in Hotel A. The variance and volatility are given by:

$$s_A^2 = \frac{(0 + 1 + 1 + 0 + 4 + 4 + 0)}{(7 - 1)} = \frac{10}{6} = 1.667 \qquad s_A = 1.29$$

$$s_B^2 = \frac{(0 + 4 + 4 + 0 + 16 + 16 + 0)}{(7 - 1)} = \frac{40}{6} = 6.667 \qquad s_B = 2.58$$

The volatility of Hotel B is twice the volatility of Hotel A.

Two points: volatility does not forecast future values, only the amount of uncertainty surrounding future values; and volatility itself is subject to change as new values are added to the data series.

Risk aversion

principle 3.7

Common experience backed up by numerous research studies has shown that people value a certain outcome more highly than an uncertain outcome with the same expected value. The underlying reason is loss aversion – the fact that when there is uncertainty people are more upset by a bad outcome – a loss – than they are delighted by a favourable outcome. Because losses are in effect weighed more heavily than gains, people's subjective valuation of an uncertain outcome is lower than the expected value.

Gambling is an exception: people buy UK lottery tickets for £1 when the expected value is £0.50 (50% of income is given out in prizes) because of the possibility (however improbable) of £1m prizes. This situation is not typical of construction projects (!).

Risk aversion is also inapplicable to financial traders. They do so many deals that they are happy to take losses on individual transactions – their focus is on the overall balance of many trades. They take a risk-neutral approach.

It is now clear that the income figures in the Chauncey Club refurbishment cashflow are expected values, not certain values. This reduces Mr Black's willingness to undertake the investment. How is this built into his evaluation?

Recall from Chapter 2 that the present value of a cashflow is established by discounting the entries for future years. This is to take account of time preference. If the future incomes are also uncertain, then their present value will be further reduced because of risk aversion. Recall also that higher discount rates produce lower present values. Thus the effect of adding uncertainty to the data in a cashflow is to reduce the present value, which is equivalent to discounting at a higher rate – the *risk-adjusted discount rate*.

Referring again to Chapter 2, we saw that high discount rates lead to a short-term perspective. Therefore uncertainty, which results in higher discount rates, favours short-term decision-making.

The lower a man's income, the greater, were he rational, his aversion to risk. That is because the pain of loss is far greater for the poor man than for the rich, and the most conservative of all will be the peasant who lives at the margin of subsistence. If his gamble on something new turns out badly, he goes hungry, perhaps dies. ... Nothing more poignantly explains the reluctance of the poor farmer to try some agricultural innovation. The age-old methods have proven themselves by the fact that he is still alive.

J K Galbraith *A Life in Our Times* (1982)

Valuing uncertain investments

Mr Black wonders what risk-adjusted discount rate he should use for evaluating the Chauncey Club refurbishment alternatives. The difficulty is that risk varies from case to case, so he can't use an off-the-shelf standard value.

Fortunately for Mr Black, all investments in business or the financial markets are more or less uncertain, and therefore all investment activities reveal implicit responses to risk. The collective view of uncertainty and the risk-adjusted discount rate is embedded in the prices set by financial and property markets. By looking at the right part of the market, Mr Black can get at the consensus of investors' opinions on how to respond to risk.

It is important to look at the right part of the market, because the amount of risk and the responses to it vary widely. Mr Black is developing what is essentially a hotel facility for an oil company. Should Mr Black base his risk-adjusted discount rate on data from the hotel industry or the oil industry? The hotel industry is better, because it reflects the uncertainties that actually apply to the Chauncey Club project – regarding future room occupancy, running costs, competition from other facilities, and so on. More about this in Chapter 4.

Mr Black's first whole-life costing exercise used a 6% discount rate. As we saw in Chapter 2 (Technique 2.3), this favoured the ambitious refurbishment. Suppose that the risk-adjusted discount rate should be 8%, then the net present values of both alternatives are sharply reduced, but both remain positive and the ambitious refurbishment is still favoured. But if the risk-adjusted discount rate should be 12%, then the net present value of the ambitious refurbishment is negative and the project is not worth pursuing, whereas the net present value of the minimum

The outstanding fact is the extreme precariousness of the basis of knowledge on which our estimates of prospective yield have to be made. Our knowledge of the factors which will govern the yield of an investment some years hence is usually very slight and often negligible. If we speak frankly, we have to admit that our basis of knowledge for estimating the yield ten years hence of a railway, a copper mine, a textile factory, the goodwill of a patent medicine, an Atlantic liner, a building in the City of London amounts to very little and sometimes nothing; or even five years hence. In fact, those who seriously attempt to make any such estimate are often so much in the minority that their behaviour does not govern the market.

J M Keynes *The General Theory of Employment, Interest and Money* (1936)

Risk-adjusted discount rates

principle 3.8

Because of risk aversion (which is really loss aversion), an uncertain income with a given expected value is less attractive to investors than a certain income with the same value. This is reflected by using a higher risk-adjusted discount rate for uncertain future incomes than for certain incomes. After discounting, the present value of the uncertain income will thus be lower than the present value of the certain income, matching investor preference.

A trade-off between risk and return is also found in the choice of investment opportunities. Risky investments will (on average) give a better return, but with the possibility of loss. The most certain investments, with no possibility of loss, give the lowest return.

refurbishment is still positive, and therefore this becomes the preferred alternative.

As we have noted before, high discount rates favour low capital cost projects, even though they may have shorter service lives and a lower level of benefits than higher capital cost projects.

Mr Black realises that the impact of risk on the discount rate is a huge factor in project evaluation.

Uncertainty and whole-life costing

The uncertainty in the Chauncey Club example related to future income. There are many other sources of uncertainty that make it extraordinarily difficult to assemble the data required by the standard method of whole-life costing. The more you think about whole-life costing, the more areas of uncertainty you identify.

Cashflow data

Standard whole-life costing requires a project cashflow for the duration of the study period. Uncertainty affects estimates of both future costs and future incomes. For example, we know today's fuel prices but there is great uncertainty about future fuel prices. Uncertainty affects the labour and materials costs for building maintenance, and indeed all factors that contribute to the project cashflow.

Building life

Virtually every building's life-history is uncertain: how will it be used, and how long will it survive before demolition? In practice demolition causes few problems for whole-life costing, as it rarely considers more than a 30-year study period, and only the most unlucky buildings are demolished so quickly (although it can happen).

In 1994 the Ionica telecomms company moved into a high profile, custom-designed headquarters building (RH Partnership architects), but in 1999 the firm went broke and the building was left empty. It had to be adapted for multi-tenancy letting.

Investor's commitment

Although uncertainty about a building's demolition is seldom a problem for whole-life costing, the duration of the owner's commitment can be crucial. When the average life of a fashion boutique is about five years, and those that survive re-brand themselves frequently, there is very little point doing a 20-year whole-life costing study of a boutique fit-out. In many businesses and institutions unpredictable change is the norm. A large firm can build a new headquarters, get taken over (or go broke), and vacate the HQ building in a few years.

Building components are often replaced before they are worn out, especially in fashion-driven sectors like retailing. It would be wasteful to invest in very long-lasting components for shopfitting in a fashion boutique.

Component service life

The service life of replaceable components in a building is uncertain. Components are replaced when they are broken or worn out, but the threshold of being 'worn out' is itself uncertain. Research into component life is valuable, but it can only reduce and not eliminate uncertainty, since actual service life is the result of many factors including heavy or light usage, degree of exposure or pollution, standard of maintenance, quality of manufacture, accidental damage. A great deal of important research goes into reducing uncertainty in service life estimates, but precise prediction is not a realistic goal.

In addition there are many cases of 'premature replacement', when components are stripped out even though they still have remaining service life, perhaps because new and better products have become available, or just because of fashion or the desire for change. These processes are inherently uncertain.

Future decisions

Standard whole-life costing assumes that building components are always replaced on a like-for-like basis. If there is a choice between, say, tiling or painting a wall, and when the building is built the wall is painted, then standard whole-life costing assumes that it will be repainted to the same specification at regular intervals throughout the study period. But in fact, if we look ahead to the time when the first paint job is due for renewal there is again a choice: tiling or painting? This is a new decision. New decisions like this are normal, not the exception.

In later years new products may become available and old ones disappear, but these processes cannot be predicted. For example, single-ply roofing was not on anyone's decision tree a generation ago, and today no-one replaces asbestos with asbestos.

It can readily be ascertained that some businesses and trades are subject to far greater risk than others; and therefore we should expect to find that the profits would be greatest in those trades that are most hazardous, so that the increased risk may thus be compensated. ... If an individual obtains ten per cent. from a copper mine when the current rate of interest is only four per cent., six per cent. may be regarded as a return for the hazardous nature of the speculation.

H Fawcett *Manual of Political Economy* (1876)

The sequence of future decisions

principle 3.9

The world view of standard whole-life costing is extremely simple – the choice between alternative components is made at the outset of the project, and is repeated on a like-for-like basis throughout the study period:

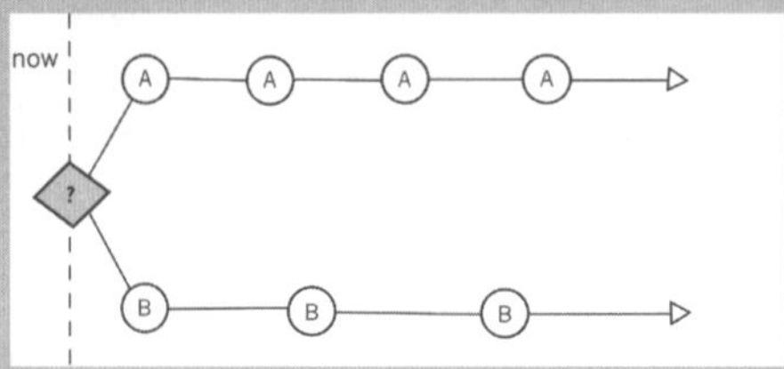

In a more realistic world view it is clear that every time you need to replace a component, a new decision between the alternatives can be made:

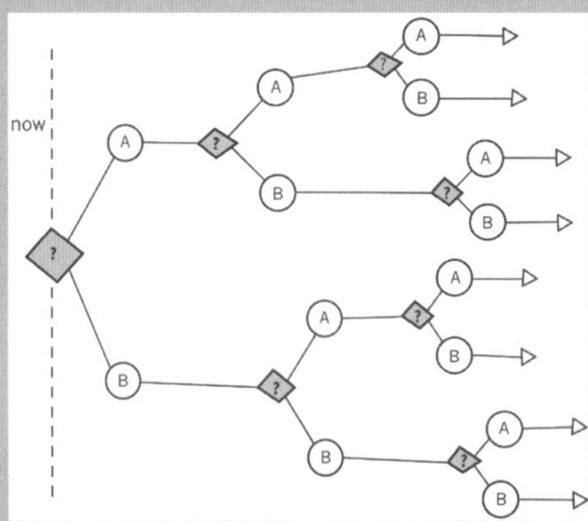

In fact, however, new alternatives are sure to appear (and old ones disappear) in the life of the building, leading to choices between alternatives that could not have been anticipated at the outset of the project:

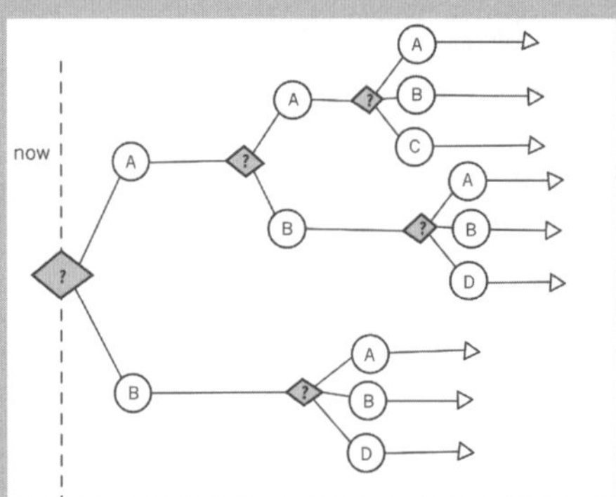

The view of the future implied by standard whole-life costing, in which today's decisions are repeated over and over again, is a drastic over-simplification.

Each aspect of uncertainty in whole-life costing tends to crank up the risk-adjusted discount rate – so favouring short-termism.

Take-up of whole-life costing

Well. Poor Mr Black – he's inclined to abandon whole-life costing completely, and he wouldn't be alone. A method of analysis built on such shaky foundations is unlikely to be adopted enthusiastically. And indeed, although standard whole-life costing has been in circulation since the 1970s its application is still far from universal, and often appears to be the exception.

We have seen that when uncertainty is added to standard whole-life costing, decision-making becomes increasingly short-term. This goes directly against the motivation for whole-life costing, which is to take a long-term view – otherwise, why bother with whole-life costing? The principle that long-term thinking is essential for long-lasting assets is so convincing that Mr Black would rather overcome the weaknesses in standard whole-life costing than abandon whole-life costing entirely. Hence the need for a new generation of whole-life costing.

4

DATA FOR WHOLE-LIFE COSTING

The conventional method of whole-life costing is mathematically straightforward, and despite shortcomings it is an important tool. To apply it effectively some fundamental issues must be understood. The key problems are establishing the cashflow data and setting the discount rate. Again, the conference centre can serve as an example.

The project cashflow

A manager wanting to run a whole-life costing analysis for a project starts with the quantification of the cashflow. Good numbers are often difficult to obtain. In particular, the decision maker needs to be very careful when using material prepared by accountants. Accountants count things in very specific and sometimes curious ways. They are not being deliberately perverse, but financial statements have to be assembled according to strict rules so that one company can be compared with another. Accountants are also very concerned about minimising taxes. The result is that their view of the world can be very different from that seen from the managerial trenches.

The following points should be borne in mind when assembling cashflow data.

Incremental costs

Only include new costs incurred by the project – exclude anything that would have arisen anyway.

Ignore sunk costs

Remember that the only important data is future cashflow: sums of money (income or expenditure) which are assigned to the future time periods when they will be received or spent. Past expenditures, or 'sunk' costs, are irrelevant. Decisions are made in the present and whatever may have happened is past and cannot be changed; over-concern about the past, as, for example, in the endowment effect, may lead to throwing good money after bad.

This principle even applies to a project that is already under way. At each stage, one has to decide whether to continue investing. The task is to determine whether further investment will get a good return, without worrying about whether the overall project including past investment is worthwhile.

When comparing 'proceed' and 'abandon' alternatives, it is essential to consider the current 'scrap' or disposal value which would be gained if the project were to be abandoned. For example, a residential development is about to start on a piece of land that was acquired years ago at agricultural prices: in the project analysis the disposal value of the land should be the current market value of housing land, not the historic purchase price. If the 'proceed' or 'abandon' comparison used the historic

The endowment effect

principle 4.1

Investors often feel loyalty to assets they have held for a long time, and are more reluctant to part with them than other assets which have been held for a shorter time. Sometimes investors are reluctant to sell an asset if its price has been falling, in the hope that it will rise and they will recover their losses.

Suppose your shares in Amalgamated Typewriters were valued at £10,000 when you inherited them from your father but have a current market value of £4,000. Selling for £4,000 feels wrong because it implies that you have given up hope of recovering the £6,000 loss. You feel that you should hold onto the shares to give them a chance to rise in value: so you are reluctant to sell. This is irrational – it is called the endowment effect.

The value of shares in Amalgamated Typewriters may rise, but if you invest £4,000 in another company whose prospects are better, perhaps Associated Microchips, then you are more likely to recover from the loss by switching. Loyalty in this situation is irrational – it is much better to forget the past performance and focus on future expectations.

Past performance may be a valuable guide to future expectations, but it is only the future that counts.

When deciding whether to hold onto an investment or switch, include the cost of switching (banker's or broker's charges, tax implications, etc) and any tangible benefits that come from the asset you currently own, for example free servicing of your typewriters(!).

The endowment effect is typical of inexperienced investors and is one reason why they get weaker investment returns than more experienced investors.

land value, it would underestimate the scale of the commitment to the development and make it appear more attractive than it really is.

Depreciation

One 'cost' which might appear from the accounting department is depreciation. This is irrelevant because it is not a cash expenditure. Depreciation is an accounting device for spreading the value of capital assets over longer periods corresponding to their useful life, and to reduce taxes (which *are* cash expenditures). Exclude depreciation.

Taxes

Taxes are part of the cashflow. They are an important business consideration and can make a big difference to the value of a project. Tax rules are constantly changing and vary substantially between jurisdictions, so are often the domain of specialists. With the help of expert advice, include tax implications in the project cashflow.

Overheads

Accounting departments may want to allocate a proportion of company overheads, such as head office costs, to projects. This is questionable. While overhead expenses are cash expenditures, they are unlikely to increase due to one more project. In the case of the Chauncey Club, for example, the refurbishment is not likely to add to Kingston Oil's corporate overheads, so for project decision-making the cashflow should not include overheads.

Finance costs

The project cashflow should not include interest or finance costs that are incurred by investors in the project. These payments are covered by the project returns – putting them in the cashflow would be double-counting.

Working capital

However, one cost which may not be apparent to someone with a design or real estate background is the need for working capital to run the completed project. For example, Kingston Oil's conference centre requires funds to buy and maintain an inventory of such things as linen and cleaning products, and funds to meet payrolls and to pay for food and other necessities before the receipt of income from guests. This working capital is a cash investment into the project.

Inflation

In his cashflow projections for the Chauncey Club refurbishment, Mr Black simply slotted in the numbers provided by Peter Williams, the architect, for maintenance costs and by Mrs Van der Vorst, the conference centre manager, for post-refurbishment incomes. They used today's prices. But what about inflation? If lift servicing, for example, costs £500 per year now, exactly the same job is likely to cost more in 10 years time. Does this make a nonsense of the calculations?

No. Inflation doesn't change the inherent value of products and services. With inflation, cash prices will be higher next year than this year, and more the year after, and so on; these are called current prices or *nominal prices*. The increases due to inflation are measured by an inflation index: you can use the index to turn next year's or the year after's inflated cash prices back into today's non-inflated prices; these are called *real prices*.

A cashflow with incomes and expenditures in nominal prices (rising with inflation) can therefore be indexed into a cashflow in real prices (inflationary growth stripped out); this has no effect on the inherent value of the incomes and expenditures.

Inflation: real prices and nominal prices

principle 4.2

Inflation is tracked by indices, such as the retail price index, building tender price index, etc. For example, the UK tender price index was set at 100 in 1976 and by 2000 it had risen to 350: a building project costing £100,000 at 1976 prices would cost £350,000 at 2000 prices, even though the real value of the construction work is identical in both cases.

Because of inflation, construction costs are usually 'date stamped' to the quarter-year: a typical cost report would say that a project cost '£4.1m (Q4 2004)'. By using the dates and the inflation index, prices from different periods can be compared.

Inflation

technique 4.1

If there is a steady rate of inflation, the nominal cost of a given product or service increases every year in a predictable way. With inflation rate h, the nominal cost increases between years i and i+1 in the following way:

$$n_{i+1} = n_i \times (1 + h), \text{ or } n_i = \frac{n_i + 1}{(1 + h)}$$

Assume that the real cost (v) is equal to the nominal cost in Year 0 (n_0), and stays the same. Then the real cost (v) and nominal cost in the ith year (n_i) are related like this:

$$v = n_i = v \times (1 + h)^i, \text{ or } n_0 = \frac{n_i}{(1 + h)^i}$$

By applying this to each entry in a nominal cashflow you can derive the corresponding real cashflow. To get the present value of a nominal cashflow (n_0, n_1, n_2, ... n_i...) you discount using the inflation rate (h) *and* the real discount rate *(r)*:

$$\text{NPV} = n_0 + \frac{n_1}{(1 + r)\times(1 + h)} + \frac{n_2}{[(1 + r)\times(1 + h)]^2} + \ldots \frac{n_i}{[(1 + r)\times(1 + h)]^i} + \ldots$$

A worked example is shown below, where expected inflation is 2.5% per year and the real discount rate is 6% per year. The net present value, which is always expressed in today's prices, is the same for both real and nominal cashflows. Note that to discount by 2.5% *and* 6% the rates must be multiplied together (*not* added). Multiplying the rates gives a nominal discount rate of 8.65% (1.06 × 1.025 = 1.0865).

Year	*Real cashflow*	*Real cashflow discounted by 6% per year*
0	–£750,000	–£750,000
1	£137,000	£129,245
2	£137,000	£121,930
3	£137,000	£115,028
4	£137,000	£108,517
5	£137,000	£102,374
6	£137,000	£96,580
7	£137,000	£91,113
8	£137,000	£85,955
9	£137,000	£81,090
10	£137,000	£76,500
Net Present Value		£258,332

Year	*Nominal cashflow inflated by 2.5% per year*	*Nominal cashflow discounted by 6% and 2.5% per year*
0	–£750,000	–£750,000
1	£140,425	£129,245
2	£143,936	£121,930
3	£147,534	£115,028
4	£151,222	£108,517
5	£155,003	£102,374
6	£158,878	£96,580
7	£162,850	£91,113
8	£166,921	£85,955
9	£171,094	£81,090
10	£175,372	£76,500
Net Present Value		£258,332

When the cashflow is in real prices it can be discounted with the real discount rate to get the present values. If the cashflow is in nominal terms it needs a nominal discount rate, which includes a factor reflecting the market's perception of future inflation. It doesn't matter whether real or nominal prices and discount rates are used, because they give the same present value. But real and nominal figures must never be mixed.

There are strong arguments for using real prices for construction project cashflows. One is that future rates of inflation are hard to predict. You can attempt to model future inflation and incorporate it into a model, but the only thing you can know about a model of future inflation is that it will be wrong. Uplifting future prices with a guesstimated inflation rate adds nothing compared to working with real prices.

Therefore Mr Black's anaylsis is not invalidated by future inflation. It is safe to take the figures from Peter Williams and Mrs Van der Vorst and use them in a real cashflow, but Mr Black must remember to use a real discount rate.

Cashflows which increase or decrease over time

A project cashflow may be subject to an ongoing increase or decrease in real terms, not just due to inflation. For example, in the case of the Chauncey Club one might believe that overall demand for conference accommodation in the area will increase, allowing room charges to rise at a faster rate than inflation; or that a shortage of skilled manpower will cause building maintenance costs to rise faster than inflation.

In these situations, the cashflow projections can be based on a regular rate of growth or decline. This can be dangerous, so caution must be observed: it is highly unlikely that real growth or decline will ever persist over a period of decades. If, for example, the conference centre revenue forecast was to grow at a few percent per year in real terms, it would eventually exceed the size of the global economy.

Cashflows that increase or decrease over time

technique 4.2

In some situations it may be anticipated that the project cashflow, or part of it, will rise or fall in real terms (this is different from price increases due to inflation). This might be the cost of maintenance contracts for such items as lifts, flat roofs, mechanical services, etc.

Suppose that the cashflow in the first year (b_0) will increase at a constant rate (g), then the cost after one year will be:

$$b_i = b_0 \times (1 + g)$$

and the cost at Year n will be:

$$b_n = b_0 \times (1 + g)^n$$

The numbers generated by this formula can be included in each year of the project cashflow.

The formula produces remarkable results if it is left to run for a long period. Suppose the maintenance cost at the outset is £100,000 per year and it is expected to grow at 5% per year in real terms, then after 30 years the cost will be £432,000 per year – and this is real growth without inflation.

The question of whether construction costs have risen in real terms over the last thirty years can be investigated by comparing retail price and building cost indices.

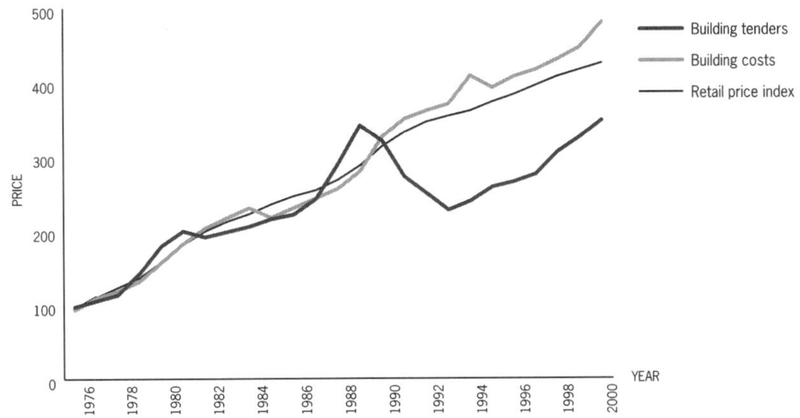

This shows that in the UK over the period 1976–2000, a period of dramatic social and economic change, construction costs kept closely in step with the economy as a whole, as represented by the retail price index. On the other hand, tender prices, which are more sensitive to the level of demand for construction, rose in real terms in the economic boom of the late 1980s but crashed in real terms in the depression of the early 1990s. This data does not support the common view that construction is becoming relatively more expensive than other industries (although it has not matched the falling costs of electronics and manufactures from developing countries). The data does not encourage the idea of predicting a steady rise in costs over a long period of time.

Market forces tend to restrict unlimited growth. If conference centre profits rise, more centres will be built, thereby depressing rates and occupancy levels. Similarly, shrinking businesses may not decline to extinction but retain a base level of activity: after all, there are still companies that manufacture buggy whips and typewriter ribbons.

Choosing the discount rate

Having made the best forecast for the project cashflow, the next challenge is to choose the discount rate for calculating the present value of the cashflow. This is the biggest single issue in whole-life costing: the number that is chosen has an immense impact on the results of any whole-life costing exercise.

We call discount rate a single issue, but one reason why it is so difficult to select is that it encapsulates in one number a wide variety of factors. It is actually a composite.

The good news is that it depends on factors connected with the project. If the project is well enough understood, it should be possible to figure out the right discount rate: it is not a mystery ingredient that appears from nowhere.

Everybody involved in a construction or real estate project is affected by the discount rate, because it determines which projects go ahead and what specifications are selected. Although everyone does not have to participate in the sometimes arcane discussion which surrounds the setting of the rate, an understanding of what is going on and whether valid assumptions are being made will allow all participants to contribute effectively and minimise the risk of under- or over-investment.

Ingredients of the discount rate

The discount rate reflects three factors:

- inflation
- time preference
- risk

Inflation was discussed earlier in this chapter: you can include it or exclude from your analysis, but never mix the two approaches.

Time preference was discussed in Chapter 2. It is a background consideration which applies across the board. It is unusual for time preference issues to play a part in setting the discount rate for a specific project.

Risk, on the other hand, does not affect all projects in similar ways. Different investment opportunities have very different levels of risk. When undertaking a project analysis, understanding risk and translating it into a suitable discount rate is all-important.

Risk and the discount rate

In practice, the discount rate for investment in a construction project is mainly determined by the amount of uncertainty or risk. The fundamental connection between investment and risk was explained in Chapter 3: because of risk aversion (or more properly, loss aversion), investors require a higher return from high-risk projects

Opportunity cost

principle 4.3

When an investor chooses between alternatives, a penalty of making a choice is that the other opportunities are foregone. For example, if a piece of land is used for a retail park it cannot be used for a housing development: the benefit that would have resulted from the housing use is an *opportunity cost*.

Each of the alternatives to the one selected has an opportunity cost, but it is only the highest one that has to be considered: the benefit of the highest alternative use. If this opportunity cost is higher than the benefit of the alternative selected, then the shortfall should be added to the alternative selected to establish its true cost. For example, suppose that the owner of an elegant villa in the suburbs of an expanding city is offered £5m for the site: he has two alternatives – stay in the villa and forego £5m, or sell for £5m. There is no financial outlay attached to staying, but it has an opportunity cost of £5m, so the true cost of staying is £5m.

When an investor chooses from a range of widely differing alternatives – government securities, a construction project, or the proverbial used car business – the wisdom of the choice is not evaluated on some absolute scale of values, but by reference to the opportunity costs of the alternatives.

Choosing the alternative with benefits that are higher than the opportunity costs of all other alternatives, ensures that resources are invested in the most efficient way.

than low-risk projects, so a higher discount rate is used for calculating the present value of a high-risk cashflow than a low-risk cashflow.

Potential investors compare, often unconsciously, the risk and returns of a building project with the risk and returns of other opportunities, such as government securities, emerging market stocks, or a brother-in-law's used car business. Each has a level of risk and an appropriate discount rate – as the level of risk rises the discount rate rises too.

Where a project has no risk, the *risk-free rate* can be used as the discount rate.

Where there is risk, a *risk premium* is added to the risk-free rate, giving the *risk-adjusted discount rate*. Generally speaking, the risk premium should be proportional to the amount of risk: if investment A has twice the risk of investment B, the risk premium for A is twice the risk premium for B.

A construction project encompasses a multitude of investment decisions for different components, which have widely varying amounts of risk. Whole-life costing for building components should use appropriate discount rates. This is done implicitly by experienced managers, who might, on the same project, require a very short 'payback period' for innovative energy-saving plant (which equates to a high discount rate), but pay extra for clay

Risk-free rate, risk premium and risk-adjusted discount rate

principle 4.4

The *risk-free rate* is the return that can be obtained from an investment with the least possible risk. This is usually taken to be an investment in government securities issued by a stable and financially sound government.

The risk-free rate varies over time and from place to place, but at a given time and place it is relatively easy to establish. In this book we take the risk-free rate to be 4.5% in nominal terms (no adjustment for inflation), or 3% in real terms (inflationary growth stripped out).

Where an investment has risk, a higher discount rate is appropriate, so a *risk premium* is added to the risk-free rate. The greater the amount of risk, the higher the risk premium.

The risk-free rate plus the risk premium gives the *risk-adjusted discount rate*.

tiles with a 60-year life rather than concrete tiles with a 30-year life (which equates to a low discount rate): this is because the expected benefits from innovative plant are risky, but there are minimal risks associated with the long-term performance of clay tiles. An average discount rate for the project would not distinguish the component-specific amounts of risk.

The principle that the risk premium should be proportional to the amount of risk raises two questions:

How is the amount of risk quantified?
Where historic data is available, past volatility is used to quantify risk: twice the volatility means twice the amount of risk. (A risk-free investment has zero volatility.)

Where uncertainty is quantified with a binomial tree, the difference between high and low values is a measure of risk: twice the difference means twice the amount of risk. (The 'binomial tree' for a risk-free investment is a single line – the difference between high and low values is zero.)

How is the amount of risk translated into the risk premium?
This cannot be done theoretically – it requires empirical data. One has to have data about a comparable benchmark where both the quantity of risk and the risk premium are known. Then one works out the ratio between the quantity of risk for the project in question and the quantity of risk in the benchmark. This ratio is applied to the benchmark risk premium, to give the risk premium for the project in question. For example, if the ratio of risk is 1.6 (the project has more risk than the benchmark), and the risk premium for the benchmark is 5%, then the risk premium for the project is 5% × 1.6 = 8%.

The risk premium is added to the risk-free rate to arrive at the risk-adjusted discount rate.

The principle is clear but the problem of getting good data makes it hard to do precise calculations. But don't be too discouraged – everybody is faced with the same problem and the best that anyone can hope for is a reasonable approximation. Someone who understands the simple principle linking risk and the risk premium is unlikely to come up with numbers for the discount rate that are completely wrong, and will be able to take advantage of better data if it becomes available.

Getting data from the market

In practice managers do not go back to first principles to work out the discount rate for a project. All the factors that have to be considered are reflected every day in the marketplace. Every project can be related to some marketplace, either closely or approximately: if there is a close connection, the discount rate can be inferred easily and reliably; the weaker the market connection, the harder it becomes to establish a suitable discount rate.

There are many kinds of market. Organised financial markets for standardised securities, with strict rules and comprehensive record-keeping, generate the best quality of data. In real estate markets unique assets are traded, with much looser rules and no obligation to disclose the deals, so data is noisier: but this is the market which counts for investment in property.

The collective wisdom of the market establishes the required rate of return on capital invested in a business venture. If the return is greater than expected, the market will respond by increasing the value of the company or project, and vice versa. The market's rate of return on a company's capital is the right figure to use as the discount rate for investments in that company's business activities. It integrates all available knowledge about the company.

Remember that markets invariably work in nominal terms (no adjustment for inflation), so discount rates established from market data will also be nominal; if cashflow data is in real terms, adjustments must be made to achieve consistency when calculating the present values.

Capital asset pricing model

For companies that are traded on stock markets, past volatility provides a measure of risk. This is how the capital asset pricing model (CAPM) establishes the rate of return corresponding to an individual company's amount of risk.

The task is to establish the company-specific risk premium to add to the risk-free rate. As its benchmark CAPM uses a portfolio of investments that includes all companies in the market: the volatility of this portfolio is called systematic or market risk; the risk premium of this market portfolio is also known. An individual company's volatility can be compared to the benchmark: the ratio of company risk to market risk is the company's unique risk, called its β (beta). The company's risk premium can then be established: it is β times the market portfolio's risk premium.

Systematic or market risk is shared by everyone – it is due to the upward or downward movement of the economy or the market as a whole (tracked, for example, by national GDP figures or the Dow Jones or FT 100 index).

Unique risk is due to features of a particular company, such as the sector of the economy in which it operates or its managerial capability. CAPM estimates this by

Capital asset pricing model (CAPM)

technique 4.3

The capital asset pricing formula establishes the risk-adjusted return (r_i) for a particular asset (i) by comparing it with the risk-free rate (r_f) and the average return for the market as a whole (r_m). The essential piece of data is the asset's β (beta), which is its volatility compared to volatility of the market as a whole. This value (β_i) is used as a multiplier which is applied to the market risk premium ($r_m - r_f$), to give the particular asset's risk premium.

The basic single period CAPM relationship giving the asset's risk-adjusted return is:

$$r_i = r_f + \beta_i \times (r_m - r_f)$$

For the Chauncey Club conference centre example, the inputs for the CAPM calculation can be drawn from market data. The risk-free rate is 4.5%, the expected overall market return is 11.5%. The β of equity in conference centres is 1.2, because they are more volatile than the market as a whole. Like all market figures, the rates are nominal, not real. Thus the yield that the market expects for investment in the Chauncey Club (r_C) is given by:

$$r_C = 0.045 + 1.2 \times (0.115 - 0.045) = 0.045 + 1.2 \times 0.07 = 0.045 + 0.084$$

$$= 0.129 = 12.9\%$$

The CAPM calculation suggests that shareholders in a company in the hospitality industry expect a nominal rate of return of 12.9%.

comparing how an investment opportunity in the company or industry moves up or down in comparison to the movement in an overall market index: it may be less risky than the market as a whole (β is less than 1), or more risky (β is more than 1). Large mature companies typically have low β values. β values for companies with traded shares are published.

How might CAPM apply to the Chauncey Club? Kingston Oil has a lower β value than companies in the UK hospitality industry. The hospitality industry has higher risk because the use of hotels is more discretionary than oil consumption: during economic downturns hotel bookings fall faster than energy consumption. Therefore the market will apply a higher risk premium for investments in hotels and conference centres than in oil companies.

A CAPM calculation suggests that shareholders in a hospitality company would require a rate of return of 12.9% (like all market rates, this is nominal and not adjusted for inflation).

Who bears the risk? – the weighted average cost of capital

Some older whole-life costing guidance suggests that the interest rate can be used as the discount rate if the project is funded with a loan, or the return demanded by investors if the project is financed by shareholders. Unfortunately this undemanding approach will almost inevitably provide an inappropriate discount rate.

Lenders, including mortgage lenders, carry a different risk from that experienced by the project as a whole: they take less risk because they typically require security and are the first people to be paid if the project or company fails; on the other hand, they do not benefit if the project is an outstanding success. Because of the lower risk, the interest rate charged by lenders will be lower than the correct discount rate for the overall project.

Shareholders on the other hand take more risk: if the project or company fails they are likely to get nothing, but if it is very successful the value of their holding will increase. So, in accordance with the principle that the risk premium is proportional to the amount of risk, the return required by shareholders will be higher than the correct discount rate for the whole project.

When a project is funded by capital from a mixture of sources with different exposure to risk, a technique for establishing project risk is the weighted average cost of capital (WACC).

Large companies know their own weighted average cost of capital. Mr Black and his colleagues working on the Chauncey Club refurbishment can get this number from Kingston Oil's finance department. They would like to use it as the discount rate for whole-life costing of the refurbishment project. However, many, if not most, real estate decisions do not meet the two primary conditions for using a company's WACC as the discount rate:

Scale: The project must represent a small addition to the overall company's activities, and its implementation will not change the market perception of the riskiness of the company.

Consistency: The project must have the same level of risk as the existing activities of the company. That effectively means an expansion of the dominant business activity of the company.

For the Chauncey Club refurbishment there is no problem with the first condition, as it is a small project being undertaken by a large multi-national company.

However, the second condition is a major impediment.

Weighted average cost of capital

principle 4.5

In calculating the weighted average cost of capital for a company or a project, all sources of capital are included, such as debt, equity, bank loans, and off-balance sheet forms of financing like leases. Debt and equity are often the most important. The returns required for each source of capital are averaged in proportion to their contribution to the total capital, giving a single rate of return – the weighted average cost of capital.

Weighted average cost of capital (WACC) technique 4.4

If a project is funded from more than one source of capital, and the sources require different rates of return, the overall rate of return that must be achieved by the project to satisfy its investors is given by the weighted average cost of capital.

Suppose the proportions of capital from different sources are:

$$p_1, p_2, p_3, \ldots p_n, \text{ where } p_1 + p_2 + p_3 + \ldots p_n = 1$$

and the required rates of return are:

$$r_1, r_2, r_3, \ldots r_n$$

then the weighted average cost of capital is given by:

$$\text{WACC} = (p_1 \times r_1) + (p_2 \times r_2) + (p_3 \times r_3) + \ldots (p_n \times r_n)$$

Simplifying, we can eliminate the more complex financial vehicles and include just two sources of capital, debt (or loans) and equity (or shareholding), Suppose the total capital is C, where D is provided by debt and E by equity, and the rates or return are r_D and r_E, then:

$$p_D = \frac{D}{C} \quad \text{and} \quad p_E = \frac{E}{C}\text{, where } D + E = C.$$

$$\text{WACC} = (p_D \times r_D) + (p_E \times r_E)$$

In the case of the Chauncey Club refurbishment, using hospitality industry figures we can estimate that 40% of the capital is provided by debt and 60% by equity, and that the market requires a 6.9% return on debt and 12.9% on equity. It is then possible to calculate a reasonable cost of capital to use when considering investments in the hospitality industry:

$$\text{WACC} = (0.4 \times 0.069) + (0.6 \times 0.129) = (0.0276 + 0.0774) = 0.105 = 10.5\%$$

This is derived from market data, so it is a nominal rate. Note that the WACC is different from the return required by either debtholders or shareholders.

The project not the organisation

Large organisations make many investments, with varying risk and return profiles. In project evaluation it is necessary to use a discount rate which reflects the project in question. An average figure for the organisation is unlikely to be correct, so both the CAPM and WACC methods have to be used with caution as they are geared to establishing the company-wide risk premium, although both methods can also be applied at project scale.

Where a property or real estate investment is different from the company's main business, the company's risk-adjusted discount rate cannot be adopted. In the case of the Chauncey Club, the conference centre is very different from Kingston Oil's main energy business, and is likely to have a different risk profile.

This makes sense. The riskiness of the Chauncey Club project would be the same whether the investor was an established oil company (relatively low risk) or an innovative high-tech firm (relatively high risk), so using the investing company's discount rate would distort the analysis of the conference centre.

Consider also that Kingston Oil has an alternative to owning the conference centre: it could sell it and send its employees to locations under other ownership. The benefit that the oil company receives from having its own conference facility is the stream of savings obtained by not paying for outside resources. If conference centre charges are high the saving is substantial, but if they are low the saving is low. Hence, it can be seen that the return varies with conditions in the hospitality industry, not those in the energy industry. Therefore, the decision-maker should look to the level of returns that the market demands for investment in the conference and hotel industry.

Thus a company should demand returns that are higher or lower than its company-wide discount rate for diversifications outside its existing main business. This also applies to diversifications by real estate companies. A company holding only hotels would be regarded differently in the marketplace from one holding only office buildings, and the discount rate of a company specialising in hotels might not be appropriate for a venture into offices.

Analysis of comparables

When the CAPM and WACC methods cannot be used, the market can still provide useful data from comparable companies or projects. The comparables must be drawn from a market in which data of a reasonable quality and quantity can be obtained, which excludes the brother-in-law's used car business. Typical sources include:

Traded stocks: Stockholders share the ownership of a company, and its profits. Good quality data is available for the prices, yields and volatility of traded stocks.

Traded bonds: Bonds are issued in exchange for long-term loans to companies and government bodies; bondholders are paid interest and have a guaranteed repayment date. Good data is available, although many bonds have special features which affect how they are valued by the market.

Real estate sales data (such as it is): How the marketplace behaves when it buys or sells property assets – how much does it pay for a stream of expected incomes? The quality of this data is much more patchy and 'noisy' than for stocks and bonds.

For the Chauncey Club management team, estate agents' reports give some insights, although care must be taken to compare like with like. Hotels include up-market luxury facilities as well as very basic structures without additional amenities. These facilities serve different clientele, with different risk profiles, so incomes and asset prices may move differently. Available data from a variety of recent estate agents' reports on sales of over 6,000 hotel rooms in the United Kingdom suggested that hotels (outside London) traded at an average price of just under £58,000 per room. The Chauncey Club is mid-range in terms of both quality and location, so the team concludes that its rooms might trade for roughly that amount. This is a useful indicator: it would be unwise to end up paying more to refurbish the accommodation than it would fetch on the open market.

A second way of using market data is to compare the current returns of similar facilities that have been traded in the marketplace. Again, such numbers can be found in the reports of estate agents, covering the sales of thousands of units. The usual property limitations hold true: each piece of real estate is different, appealing to different markets, and sold under different conditions, with different contract terms. With this proviso, in the last couple of years current returns for hotel sales ranged from about 7.5% to 11% (in nominal terms).

Sources of difficulties with market data

Caution must be exercised when using comparables to set the discount rate for a project: companies in the same sector can vary widely, not to mention the individual projects under consideration.

The analysis of real estate data is far more difficult than assessing portfolios of stocks and bonds:

Stocks are interchangeable: If there is a strong demand for General Motors stocks, any owner can sell to any buyer. But all property assets are unique so the connection between supply and demand is more complex.

Slow market response: When dealing with traded securities, money can move very quickly to balance supply and demand and cause an adjustment to market prices. In property transactions there are long time delays between agreements and completions, and high transaction costs.

Market data: Detailed minute-by-minute statistics are readily available for stock and bond markets, unlike property markets where even aggregated quarterly data is rare.

These difficulties cause uncertainty about the discount rate for a property investment. This uncertainty will itself tend to push up the discount rate, which in turn leads to a short-term perspective – contrary to the spirit of whole-life costing which seeks the long-term view: therefore, better data about the real estate market is an ally of whole-life costing.

A note on the public sector

The use of market data for assessing risk and the risk-adjusted discount rate is directly applicable to private

Transactions Costs

principle 4.6

These are the costs incurred when buying or selling anything, including stocks, bonds – or real estate. The costs may include broker and legal fees, moving or transfer costs, and taxes. These vary widely from place to place: in some emerging economies selling property can involve numerous government fees and approvals, and take up to a year.

sector investment. What about public sector investment? The same underlying principles are applicable, and should be followed in the evaluation of public sector projects to avoid under- or over-investment. The UK Treasury's guidance on public sector investment states that project costs and benefits should be estimated on the basis of opportunity costs.

Difficult to choose a discount rate? The internal rate of return

Given the difficulties in identifying the appropriate discount rate, it is tempting to reverse the process, determining the discount rate at which the project cashflow has a net present value of zero. This discount rate is called the internal rate of return (IRR).

Generally projects with a high IRR are preferred. A high IRR indicates a cashflow with lots of positive numbers which have to be heavily discounted to get to a zero net present value. A low IRR suggests that there is a balance of positives and negatives in the cashflow. A cashflow with a negative cash balance usually has a negative IRR.

Unfortunately, the project IRR has to be interpreted with caution:

IRR ignores risk: We know that risky projects must give high returns, so these projects tend to generate a high value for the IRR. If investment projects are selected on the basis of the highest IRR, risky projects will be favoured, but we have seen that the analysis of risk is one of the most important aspects of investment evaluation. In contrast, the net present value calculation for cashflows with different levels of risk will use different risk-adjusted discount rates.

Cash balances: The IRR assumes that any cash balances held by the project earn income at the same rate as the IRR – but the IRR is an arbitrary number which bears no relation to investment opportunities. In contrast, calculations with a realistic discount rate make the more reasonable assumption that cash is reinvested at the discount rate.

Internal rate of return

technique 4.5

The internal rate of return (IRR) of a cashflow is the discount rate for which the net present value is 0. For a cashflow v_0, v_1, v_2 ... v_n the internal rate of return (z) produces the result:

$$v_0 + \frac{v_1}{(1+z)} + \frac{v_2}{(1+z)^2} + \ldots \frac{v_n}{(1+z)^n} = 0$$

IRR is used to compare investment alternatives. Generally, the alternative with the highest IRR is preferred, assuming equal levels of risk.

For example, consider an office building of 5,000m^2 that can be fitted out to a higher grade for £1,000/m^2 or to a lower grade for £800/m^2. The higher grade can be rented for £200/m^2 per year for 15 years, and the lower grade for £180/m^2, with further £250,000 refurbishments in years 6 and 11. The cashflows are shown below:

Year	*Higher grade*	*Lower grade*
0	–£5,000,000	–£4,000,000
1	£1,000,000	£900,000
2	£1,000,000	£900,000
3	£1,000,000	£900,000
4	£1,000,000	£900,000
5	£1,000,000	£900,000
6	£1,000,000	£650,000
7	£1,000,000	£900,000
8	£1,000,000	£900,000
9	£1,000,000	£900,000
10	£1,000,000	£900,000
11	£1,000,000	£650,000
12	£1,000,000	£900,000
13	£1,000,000	£900,000
14	£1,000,000	£900,000
15	£1,000,000	£900,000
Internal rate of return (IRR)	18.4%	20.6%
'Payback period' (years)	5	4.4
NPV at 10% discount rate	£2.6m	£2.6m

The IRRs for the two alternatives are 18.4% and 21.3%. On this basis the lower grade refurbishment is preferred. The ranking of alternatives is the same as given by the simple payback method. This is usually the case in construction projects.

These are high discount rates and therefore give greater weight to early entries in the cashflow – a short-term view. However, for lower discount rates, say 10%, the NPVs of the alternatives are equal as shown in the table. If the discount rate is lower than 10%, the higher grade fit-out will have a higher NPV.

The NPV evaluation method with a realistic discount rate is preferable to the IRR method.

Multiple solutions: Some cashflows have more than one value for the IRR. It does not happen in real estate situations where an initial expenditure is followed by a series of incomes. However, it can occur if there are large negative entries some way in the future. This may apply in cases such as restaurants and fashion-oriented shops with frequent and substantial refurbishment expenditures. Another case would be a project with a massive terminal cost, perhaps to de-commission a nuclear power station or rehabilitate a contaminated site.

No solutions: Construction industry whole-life costing often compares the first cost and running costs of alternative building components, where the cashflow has no figures for income. A cashflow of this type will not have an IRR.

IRR is sometimes useful for a quick comparison of cashflows, if the alternatives have similar amounts of risk. It is better than the simple payback method, as it takes account of service life beyond the end of the 'payback period'.

Summing up

We are straying beyond the needs of Horatio Black or most managers in property and real estate. One thing should be clear: setting the correct discount rate for a project requires careful analysis with particular attention to the level of risk. Managers in property and real estate sometimes rely on others to tell them what rate to use, but they should also satisfy themselves that the rate is credible, so that they can have confidence in the results of their whole-life costing exercises.

The market comparisons tell Mr Black the risk-adjusted discount rate for a hospitality project is around 10% to 10.5%. This is a nominal rate, anticipating inflationary growth in the stream of incomes; but his cashflow is expressed in real prices, without inflation. The real cashflow must be discounted with a real discount rate, and to get the real discount rate from the nominal rate he must deduct something for inflation, which is currently about 2% per year. He decides that a real risk-adjusted discount rate of 8% is a reasonable estimate for evaluating the Chauncey Club refurbishment.

With the cashflow data that was set out in Chapter 2 and a discount rate of 8%, the net present value for the minimum refurbishment is £169,300 and for the ambitious refurbishment it is £255,100. Both are positive, so they are worthwhile projects, but the NPV of the ambitious project is higher so it is preferred. Mr Black can take this recommendation to the board of Kingston Oil with a clear conscience.

However, Mr Black knows that the analysis that led to this recommendation relies on some doubtful assumptions, and he is now sufficiently interested in whole-life costing to ask how things could be done better. The following chapters make some suggestions. We will visit the Chauncey Club just once more, and then move on to new case studies.

5

LIFECYCLE OPTIONS

The future is uncertain so decisions made today may turn out to be ill-adapted to future circumstances. Rather than making decisions today, it would be better if they could be made in the future, in response to unfolding events. This is the principle of a lifecycle option – today's options allow decisions to be made in the future. This chapter explains lifecycle options and shows how they provide the basis for a new generation of whole-life costing.

The reluctant decision-maker

In the research project which led to this book a large number of managers were asked whether they used whole-life costing. Although virtually all knew about the technique, some in reasonable detail, few indicated that they ever used it. Those who did tended to use the results as only a minor part of their decision-making process. Most were happier to rely on gut feel, especially the more experienced managers. Asking why so little weight was given to whole-life costing elicited a number of responses. It seemed that most of the managers felt that the findings of whole-life costing did not seem to correspond to their experience. A number of anecdotes were told. It seemed that although whole-life costing had a certain intellectual appeal, it was too far removed from real world decision-making.

Something is clearly missing from standard whole-life costing analysis.

Making decisions later

A crucial mismatch between standard whole-life costing and people's experience is that not all decisions are made at the beginning of a project, as assumed by standard WLC. Buildings are long lived and dynamic: managers in the future can and will make decisions that affect the evolution of their buildings. This is a good thing because future managers will have up-to-date information. There is much merit in trying to avoid making decisions today: look again at Principle 3.9 on page 38. Perhaps there may after all be a Darwinian justification for the human propensity to procrastinate.

The reason for preferring to make decisions later is to avoid bad decisions, or rather, decisions which turn out to have bad outcomes. Avoiding bad outcomes is very desirable.

With future uncertainty one has to anticipate a range of possible outcomes flowing from a decision, some of which will be more favourable and some less favourable. The weighted average of all these favourable and unfavourable outcomes is the expected value of the decision. If by taking decisions later one avoids some of the unfavourable outcomes, then the expected value will be increased. This is good news, and it gets better. As the range of outcomes and the level of uncertainty attached to the decision are reduced, so the risk premium in the discount rate for project analysis will be lower.

Both these two consequences of deferred decisions, higher expected value and lower discount rate, increase the present value of a cashflow. So, relaxing the rigid assumption of standard whole-life costing, that all decisions are made in Year 0 and are irrevocable, increases whole-life value.

By neglecting opportunities for future decisions, standard whole-life costing understates the value of most construction projects, especially projects which contain many future decision possibilities.

Decisions that can be taken in the future with better information can be termed 'lifecycle options'. Rather than take the decision between alternatives in Year 0, it is often preferable to retain lifecycle options so that better decisions can be made later in the life of the building.

Buildings undergo unforeseen changes. This 1930s neo-Georgian building in London was initially the headquarters of the Foundling Hospital charity, but it became redundant and has recently been adapted and extended to create the Foundling Museum.

Finding lifecycle options

To Horatio Black, who is tenaciously seeking better tools for evaluating the alternative refurbishment strategies at the Chauncey Club, the notion of refining WLC through lifecycle options is quite exciting. Where can he find some lifecycle options?

Many lifecycle options exist but are unrecognised. In Chapter 3 we questioned the standard WLC assumption that all decisions have to be made in Year 0, and are then endlessly repeated: if in Year 0 there is a choice between painting and tiling a wall, and painting is chosen, then there is actually another choice when the wall is due for repainting (or even earlier) – it could be repainted or equally it could be tiled. This is a lifecycle option which standard WLC suppresses, thereby understating whole-life value and possibly distorting a manager's decision.

If Mr. Black can identify lifecycle options that he already has, he is making progress. Even this first insight into

Embedded and acquired options

principle 5.1

The owner of a building or property asset holds many options – the options to sell, develop, improve, refurbish, and so on. These are simply the rights that come with ownership. One reason for preferring to own rather than rent property is to have greater freedom in choosing how to use the property: the property owner holds more options than the person who rents.

These are embedded options: they come automatically with ownership, even if they are not recognised. For example, a suburban bungalow with a large garden might be sold for a higher price than its owners expected, because they did not realise that they held the option to demolish the bungalow and develop a block of flats.

Other options are acquired by some deliberate effort. For example, a parcel of land without road access to a highway cannot be developed for housing, but if its owner buys a strip of land which is wide enough to construct a road, he acquires the option to develop the landlocked parcel. The increased value of the land due to the option must exceed the price paid for the access strip, or the deal wouldn't go ahead.

Options can be destroyed as well as created. For example, if a Victorian warehouse on a city centre site is declared a historic monument and protected from demolition, the option to redevelop the site is destroyed or rendered much more difficult to exercise.

When embedded options are unrecognised they still exist, but there are two problems. First, if significant options are overlooked, property may be undervalued. Second, if options are unrecognised they may be inadvertently destroyed, reducing property value. For example, when Victoriana was out of fashion many ornate fireplaces were ripped out or boarded up; boarding up retained the option to reinstate when Victoriana came back into fashion, but ripping out destroyed this option.

Options existed long before options terminology – the terminology is intended to raise awareness and provide better tools for understanding and valuing options.

lifecycle options has an immediate attraction for many managers, in that it explains their reluctance to make long-term commitments. Up to now this has been perceived as a conflict between gut feel and the dictates of whole-life costing. However, strategies that lead to lifecycle options do have value that is missing when long-term commitments are made at the beginning of a project. For example, painting, being shorter-lived than tiling, contains the opportunity to reconsider the choice in a few years – a lifecycle option which has value and which one does not get with tiling. Standard WLC does not recognise this option or take account of its value. Gut feel often aligns with the lifecycle options approach to whole-life costing.

Recognising lifecycle options that already exist is important, but more importantly the decisions made by Mr. Black and the project team can create new lifecycle options, or they can destroy lifecycle options.

Options thinking

The concept of lifecycle options leads to new approaches to analysis and decision-making which will be discussed in the following chapters, but even before that it is important to get attuned to the idea of 'options thinking'.

The key insight is to see that real estate and the design of buildings is not just to do with physical assets, but also with the opportunities for use that are the motive for developing buildings in the first place. Over the life of the property or building only a few of the opportunities will be followed up, but since the future is unpredictable it is impossible to know with certainty which will be exploited and which ignored. Therefore, the more opportunities there are, the more likely it is that property or building will be able to respond to whatever new situations or new demands the future will bring. The opportunities for responding to future change are what we call lifecycle options. Options thinking is always on the lookout for situations where lifecycle options are needed, and for ways of providing them.

If this Victorian building in the City of London is listed, the option to redevelop will be destroyed, greatly reducing its value. The option to redevelop the site is practically the only reason that the property, which is unproductive in its present form, has any value.

From this perspective, the familiar idea of adaptability in buildings is a lifecycle option. Adaptable buildings can be reused in different ways if, or rather when, activities change over time: adaptable buildings remain useful, whereas non-adaptable buildings become obsolete.

The idea of making provision for expansion is also a lifecycle option, which can be exercised if future demand justifies it. Many types of redundancy in design create options in use: very strong floors in commercial buildings allow flexibility in the layout of machinery and storage; demountable partitions can be dismantled and re-erected in different configurations, and so on. Note that lifecycle options may never actually be exercised: that is to be expected – they give opportunities that can be exploited or not depending on how the future unfolds.

If lifecycle options can be recognised in many well-established situations and design strategies, what is the point of introducing a new set of ideas? There are two reasons. First, lifecycle options tie together a number of

Financial options and real options

principle 5.2

The lifecycle options described in this book are based on a direct analogy with financial options. Financial options have a long and controversial history, and only became fully respectable in 1973 when a formal Options Exchange opened in Chicago. This was the same year as the publication of the revolutionary Black-Scholes equation for establishing the fair price for an option. It was developed by Black, Scholes and Merton; Scholes and Merton won the Nobel Prize for Economics in 1997 (Black had died in 1995).

In a financial option, a deal is struck to buy or sell financial commodities, typically shares, at an agreed price within a specified time, but the option holder can choose whether or not to complete the transaction. The option holder pays a premium for this option contract, or option.

Naturally, the option holder exercises the option and completes the transaction if it is in his financial interest to do so (the option is 'in the money'), otherwise he lets it lapse (the option is 'out of the money'). When the option contract is drawn up, it is uncertain whether the option will turn out to be 'in the money' or 'out of the money'. If there weren't this uncertainty, nobody would bother with options.

Option contracts can themselves be traded. Financial options have become a very important way of managing uncertainty in financial markets.

When options ideas are used in business rather than financial markets, they are called real options.

An option contract gives the option holder the right to complete a transaction involving an underlying asset, and the performance of the underlying asset determines whether the option is 'in the money' or 'out of the money'. Because their value is derived from underlying assets, options are a form of financial derivative.

disparate ideas and strengthen the case for providing for future change in real estate and buildings. Second, the options approach gives access to a variety of tools for analysis and decision-making that have been developed in other disciplines, primarily the world of finance.

Decision tools are important, because acquired options usually have to be paid for. The cost is often explicit, but unless you can calculate the value of a lifecycle option you don't know whether it is worth the cost. This has been the problem up to now with strategies for adaptability in buildings: they have been 'nice ideas' which are adopted by enthusiasts and dismissed by sceptics, but rational investors have not known what to do.

Project analysis with lifecycle options

Where lifecycle options exist but are ignored, investment decisions are likely to be wrong. To illustrate the importance of lifecycle options we will revisit the Chauncey Club where Mr Black, as we have seen, has been assigned the task of deciding about the refurbishment of Kingston Oil's conference centre.

We will add one more piece of uncertainty about the future. At present Kingston Oil's retail operation has divisions in most European countries, each with an office looking after its home territory. This leads to duplication and the company has decided that in two years time it will centralise all its European retail management. The new centre might be in the UK, which would be a boost for the nearby Chauncey Club, or it might be in Belgium, which would take existing business away from the Chauncey Club. Today the outcomes are balanced 50/50. The decision will be made next year.

Mrs van der Vorst's income projection of £170,000 per year for the ambitious refurbishment at the Chauncey Club is based on the present Kingston Oil operations, including bookings from the UK retail division. If the new European retail centre is in the UK the income could be expected to increase to £220,000 per year, but if the centre is located in Belgium it could drop to £120,000. Note that average or expected value of the two income levels is £170,000

Comparing uncertain outcomes

technique 5.1

Three Year 0 decision alternatives for the Chauncey Club refurbishment have to be compared, when there is still uncertainty about the location of the European retail centre which will not be known until Year 1. For each alternative the binomial tree of possible future outcomes can be drawn; where there are two possible outcomes the binomial tree branches; where there is only one possible outcome a single line is drawn.

The first alternative is to abandon the refurbishment; there is no investment and no benefits. The present value of this investment is 0.

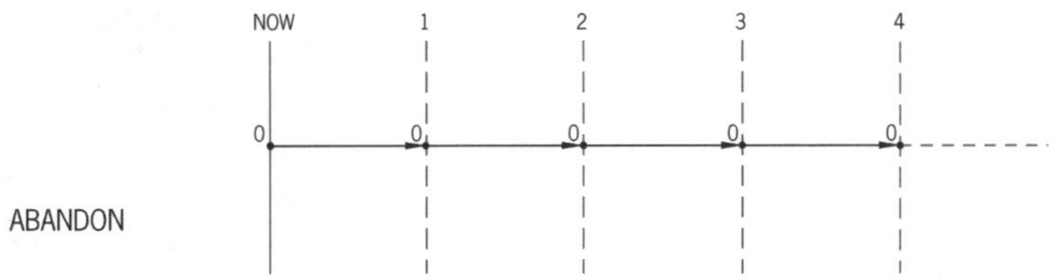

The second alternative is to refurbish in Year 0. The investment of £1.2m is made in Year 0 and a year's trading under existing conditions generates an income of £170,000 in Year 1. From Year 2 onwards there are two possible outcomes, one with high income of £220,000 per year and one with low income of £120,000 per year, depending on the location of the retail centre. The two outcomes are equally probable, so the probability of 0.5 is marked on the branches in the binomial tree. With 15 years income and a real discount rate of 8%, the present value of this investment is £255,100.

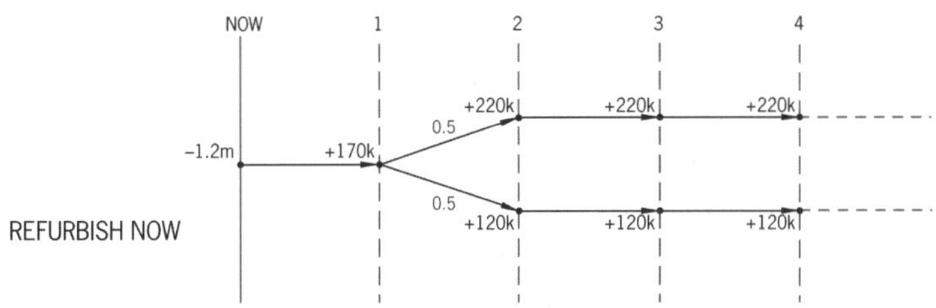

The third alternative is to retain the option to refurbish. No investment is made in Year 0. If the retail centre is located in the UK, the refurbishment is carried out in Year 1 and generates a high income for 15 years from Year 2. If the retail centre is not located in the UK, the refurbishment is abandoned. Again, the two outcomes are equally probable, so the probability of 0.5 is marked on the branches in the binomial tree. With a real discount rate of 8%, the present value of this investment is £316,200.

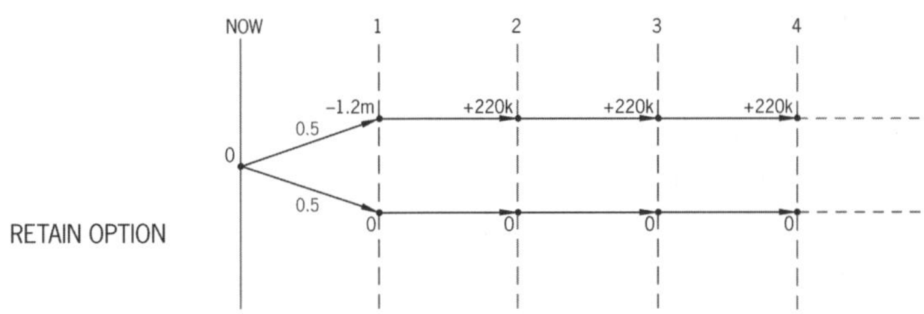

Retaining the option to refurbish is the most attractive of the three alternatives.

per year, and if this expected value is used for calculating the present value of the refurbishment the result is the same as for an income of exactly £170,000 per year. As we have seen, that gives a net present value of £255,100 for a 15-year service life, with a refurbishment cost of £1.2m and a discount rate of 8%.

Mr Black has to decide whether to go ahead with the refurbishment or abandon the project. The positive project value suggests that he should go ahead. However, he realises that if the new retail centre is located in Belgium, Kingston Oil will regret investing at the Chauncey Club. Mr Black would much rather defer the refurbishment until he knows more about the future level of demand. But how can he resist going ahead when the refurbishment has positive project value?

In fact, Mr Black does not have to decide between two alternatives, refurbish now or abandon the project. If the refurbishment is not done this year, it can be done next year – Kingston Oil retains the option to refurbish. The option can be exercised if the retail centre is located in the UK, but not if it goes to Belgium.

Mr Black can calculate the net present value of the project if it is delayed by a year, and abandoned if the retail centre goes to Belgium: it comes to £316,200 using the 8% discount rate – more than the value of refurbishing this year. This actually understates the benefit of the option strategy, as its reduced uncertainty would justify a lower discount rate, increasing present value. Sure enough, Mr Black's hunch was correct: while there is uncertainty about the location of the retail centre, it is better to retain the option to refurbish rather than to refurbish now, because it reduces the risk of a bad outcome.

If Kingston Oil refurbishes now it faces the possibility of either a good outcome (strong demand for the refurbished conference centre) or a bad outcome (little demand for the conference centre), whereas by waiting a year Kingston Oil faces the possibility of either a good outcome (strong demand for the refurbished conference centre) or a zero-value outcome (abandoning the project). In the latter case Kingston Oil could think of selling the Chauncey Club, but that is another story.

Late in 1821 a squall took place over the contentious question of option dealings, also known as puts and calls, where a speculator would pay an agreed sum for the right to buy or sell a stock at a certain price on a given day in the future. The founding Committee [of the Stock Market, above] of 1802 had unanimously condemned such dealings as injurious ... One [petition] demanded the outright abolition of puts and calls, 'which are now so frequent as to constitute the greater part of the business done in the House and which operate materially against the interests of those who do not comply with this practice'. The other [petition] took positive pride in 'the immense jobbing which is now consequent upon Options'. ... The Committee, conscious of the extremely dubious legal standing of such dealings, sided with the first petition. ... Reaction was immediate, as members queued to express their outrage in person. ... The Committee backed down.

David Kynaston *The City of London: a World on its Own 1816-1890* (1994)

Call and put options

principle 5.3

The two basic forms of financial option are the 'call' and the 'put' – options to buy and options to sell.

A *call* option confers the right but not the obligation to buy an asset at a specified 'strike' price, within a given timescale. If the market price of the asset rises above the strike price the option is 'in the money' and is exercised; if the market price remains below the strike price the option stays 'out of the money', so will expire unexercised.

A *put* option confers the right but not the obligation to sell an asset at a specified 'strike' price, within a given timescale. If the market price of the asset drops below the strike price the option is 'in the money' and is exercised; if the market price remains above the strike price the option stays 'out of the money', so will expire unexercised.

What is the practical value of these options? Consider options to buy and sell commodities.

A jam manufacturer knows that he will have to buy 100 tons of sugar in six months. Today's market price is £120 per ton, and the manufacturer will be delighted if the price falls; but if it rises above £140 his business will be in trouble. He purchases a call option to buy 100 tons of sugar at £140 per ton in six months. This gives him the right but not the obligation to exercise the option: he will exercise it if the market price rises above £140 per ton, otherwise he will buy at the market price and let the call option lapse. The call option protects the manufacturer against the risk of a price rise.

A sugar producer will have 100 tons of sugar to sell in six months. Today's market price is £120 per ton, and the producer will be delighted if the price rises; but if it falls below £100 per ton he will be in trouble. He purchases a put option to sell 100 tons of sugar at £100 per ton in six months. This gives him the right but not the obligation to exercise the option: he will exercise it if the market price falls below £100 per ton, otherwise he will sell at the market price and let the put option lapse. The put option protects the producer against the risk of a price fall.

The options, even if they expire unexercised, reduce price uncertainty and the risk of damaging losses for both the jam manufacturer and sugar producer, and so they have value. There is an analogy with insurance: although you pay for insurance in case you need to claim, you don't feel dissatisfied if the policy expires without a claim being made.

The call and put options have to be paid for, but they cost much less than the underlying assets. The options market sets the price for any option, however complex (some financial options are amazingly ingenious).

In our example the manufacturer and producer could have agreed a fixed contract for 100 tons of sugar to be delivered in six months at, say, £115 per ton. This would have reduced uncertainty even more than the call and put options, but each party would have lost the possibility of benefiting from favourable (to themselves) market movements.

The 'refurbish now' and 'retain option' alternatives both consist of a bundle of two possible outcomes: the first bundle has good and bad outcomes, the second has good and zero-value outcomes. The second is clearly a more attractive bundle.

In many cases, the lifecycle options approach accords better with experienced managers' gut feel than project analysis that excludes options. But don't assume that it is always better to hold an option than to proceed with or abandon an investment project: this won't always be true. By identifying and valuing lifecycle options, decisions under uncertainty can be made more rationally – a tremendous advance compared to the agonised tussle between gut feel and over-simplified project analysis.

Characteristics of lifecycle options

All lifecycle options reduce uncertainty and increase project value, because they enhance expected value and reduce the risk premium in the discount rate. When it comes to putting a value to a specific lifecycle option the following characteristics have to be considered.

The amount of uncertainty

In a situation with no uncertainty about the future, lifecycle options are pointless – there is no difficulty about making all decisions in Year 0. As the amount of uncertainty about the future increases, the value of lifecycle options increases as well. Options rely on uncertainty.

Duration of the option

Some lifecycle options are effectively perpetual, like a property owner's option to sell. Others have a fixed term; for example, planning consents in the UK are valid for five years, giving the property owner the option to undertake the approved work within that period – if the option is not exercised it lapses. The longer an option lasts, the more chances there are that the holder will want to exercise it. Therefore the longer the life of an option, the higher its value.

Nick Leeson and other disasters

principle 5.4

Mention the word 'option' and a series of financial disasters come to mind. It is worth stopping to consider some of them, to dispel the notion that involvement with options and other derivatives is a certain route to disaster.

Nick Leeson's adventures in derivatives, losing about one billion US dollars and causing the 1995 collapse of Barings Bank (est. 1763), are entertainingly presented in his book Rogue Trader and the movie of the same name. He was trading short-term options whose main role is to manage uncertainty, but as the losses piled up he got involved in increasingly risky deals (high risk can bring high returns) in the hope of balancing the books – but the losses just kept mounting. The real failure was the bank's astounding lack of control over staff activities.

More interesting if less entertaining is the case of Long Term Capital Management, an investment fund operating with the benefit of Nobel Prizewinners Myron Scholes and Robert Merton. The idea was that new understanding of financial markets would allow uncertainty and risk to be overcome by intricately balanced investment strategies, guaranteeing good returns. Everything went well and money poured into the fund, until the 1998 default by the Russian government which led to worldwide financial turmoil. LTCM's investment strategy would probably have worked in the long term but became untenable in the short term: their models had not anticipated the risk of such a severe and sudden dislocation of the markets. They managed to lose over four billion US dollars in a few weeks.

Despite lurid counter-examples such as these, when used prudently derivatives are a respectable and vital part of the financial system that we all rely on. They certainly can't be un-invented.

The trigger point for exercising the option

The question of whether or not a lifecycle option is exercised depends on the way the future evolves. Every option has a trigger point, and if this point is reached the option is 'in the money' and will be exercised. The trigger point is linked to the source of uncertainty which creates the option value. For example, in the Chauncey Club example earlier in this chapter the trigger point for undertaking the refurbishment would be Kingston Oil's decision to locate its European retail centre in the UK. In the example there was a 50% probability of this trigger point being reached; but if there had only been a 25% probability of a UK location, there would have been less chance of reaching the trigger point and getting 'into the money', so the option to refurbish would have been less valuable. Options with an easily reachable trigger point are more valuable that options with a more remote trigger point.

The cost of exercising the option

Some options can be exercised at no cost. For example, if a building is repainted every five years, there is a no-cost option to change the colour every five years. Other options incur differing cost penalties when they are exercised: for example, consider a framed building with the option to move non-structural partitions to create a new spatial configuration; if the partitions are made of plastered concrete blockwork it is slow, disruptive and expensive to exercise the option, but if they are made of demountable panels it is much quicker and cheaper to exercise the option. The lower cost of exercising the option increases its value; which is why people are prepared to pay more for demountable partitions. Thus the lower the cost of exercising a lifecycle option, the higher the value of the option.

The strategy of designing adaptable buildings is well established, as in Chamberlin Powell & Bon's extensions to the University of Leeds of the 1970s. Adaptability anticipates that the buildings may be changed by future decision-makers: in our terms, it creates lifecycle options.

The resulting benefit

The value of a lifecycle option also depends on the scale of the benefit that would be derived from exercising it. For example, the benefit derived from exercising an option to change the colour when repainting a building is not very great, so the value of the option to change colour is also modest. On the other hand, the benefit from exercising an option to extend the building stock of a new university is very great – without it the university's growth might be strangled. This is why the new universities founded in the UK in the 1960s were required to have large sites of at least 40 hectares, even though start-up student numbers were tiny: the substantial price paid for the empty land was the cost of acquiring the option to expand. Thus the value of a lifecycle option increases with the scale of the benefit that would result from exercising it.

Summary of lifecycle option value

Lifecycle options with the following characteristics have a high value:

- context of high uncertainty
- long duration, or perpetual
- easy to reach trigger point
- inexpensive to exercise
- large benefits if exercised.

Lifecycle options with the following characteristics have a low value:

- context of low uncertainty
- short duration before expiry
- difficult to reach trigger point
- expensive to exercise
- small benefits if exercised.

Many lifecycle options will exhibit a mix of these characteristics and their relative importance has to weighed when assessing option value. This can be done by gut feel or judgement, or by quantification. The following chapters present approaches to quantifying the value of lifecycle options for a variety of cases.

New generation whole-life costing

There is a vivid contrast between standard whole-life costing which requires an unrealistic prediction of the future, and the options approach which is built on an understanding of uncertainty. A depressing aspect of standard WLC is that a great deal of knowledge about future possibilities and opportunities has to be thrown away. Better investment decisions should be achievable if the entirety of our present knowledge about the future can be exploited.

The lifecycle options approach requires a forecast of future uncertainty, and the forecast may turn out to be wrong, but today's decisions have to be made using today's knowledge: it is far better to forecast uncertainty than attempt impossible predictions of specific outcomes.

Because property and real estate investments are so long lasting they virtually always involve future uncertainty. Therefore option value will almost always figure strongly in overall project value and in the choice between investment alternatives.

The application of the lifecycle options approach to whole-life costing is a new and as yet experimental departure. We believe that the new approach has immense potential, and could transform whole-life costing from a backwater into a mainstream tool for property and real estate decision-making. We feel there is justification for adopting the name *new generation whole-life costing*.

6

THE OPTION TO DEVELOP

A building owner has identified a development opportunity and must decide whether to go ahead with it. He holds a lifecycle option – the option to develop: should he exercise the option or retain it? This chapter considers this situation, and introduces techniques for options-based project evaluation.

Forgetting everything our teachers ever told us about procrastination

Harry Phillips owns an assortment of small properties acquired over a period of years. One of these is a duplex (two units, one on top of the other) that he bought almost twenty years ago on the fringes of a prestigious neighbourhood in a North American city. Containing large two-bedroom units, it has been easy to find tenants, and has commanded good rents. Over the years, Mr Phillips has replaced the kitchens and bathrooms and kept the building in reasonably good order. The building also has a development opportunity: it contains a basement, now used for storage, which could be reconstructed to provide an extra rental unit with one or two bedrooms.

Mr Phillips' suburban property. It has two apartments and a basement which could be developed to create a third rental unit.

A few years ago Mr Phillips obtained the necessary consents to allow the work to proceed, but has delayed the work, partly because of other business commitments, but also because of the returns he could obtain by investing in stocks. In early 2005 he has a sense that it may now be worth proceeding. He has enough experience to know that construction work is not easy, particularly in an occupied building, so he needs to be assured that proceeding is the correct decision.

A standard whole-life costing exercise using a reasonable discount rate shows that the net present value offered by his project is positive, indicating that it is viable to proceed. However, Mr Phillips is not fully convinced. He might just keep on procrastinating and put his money into something else, and maybe do the basement some time in the future. Recalling what his elementary school teachers told him about the dangers of procrastination, this leaves him feeling as uneasy as proceeding.

Standard whole-life costing

technique 6.1

The standard whole-life costing analysis of Mr Phillips' development starts with a cashflow setting out the construction cost in Year 0 and ten years' expected income in Years 1 to 10. Nothing has been added for inflationary growth, so this is a real cashflow.

The cashflow is discounted to get the net present value. Because the future rental income is uncertain, a risk-adjusted discount rate of 8% is used. This represents a 5% risk premium over the 3% risk-free rate. These are also real rates.

The discounting is carried out in the usual way:

$$NPV = v_0 + \frac{v_1}{(1 + r)} + \frac{v_2}{(1 + r)^2} + \ldots \frac{v_n}{(1 + r)^n}$$

where v_0, v_1, v_2 ... are the yearly entries in the cashflow and r is the discount rate.

Year	*Cash value*	*Present value*
0	-$64,000	-$64,000
1	$9,600	$8,889
2	$9,600	$8,230
3	$9,600	$7,261
4	$9,600	$7,056
5	$9,600	$6,534
6	$9,600	$6,050
7	$9,600	$5,602
8	$9,600	$5,187
9	$9,600	$4,802
10	$9,600	$4,447
Net present value		$417

The NPV is positive, indicating that this is a viable project.

The option to develop

If Mr Phillips faced a situation where he had to undertake his development this year or lose the opportunity – his approvals might be about to expire – he could go ahead without misgivings. He has good knowledge of construction and the rental market, so his cashflow forecast for the project is credible. Because he knows that future rents are uncertain he uses a risk-adjusted discount rate of 8% to get the net present value,

Mr Phillips estimates the development of the basement unit will cost $64,000 and generate an income of $9,600 income per year at current rental levels. He expects a 10-year stream of incomes before having to undertake any refurbishment or updating. The apartment and the cashflow would actually continue beyond this point, but for this exercise we are just looking at 10 years of data. The NPV of this project using standard WLC is just $416, which is fine because an NPV above zero is the test of a viable project.

Although there is a chance that rents may drop, that is covered by the risk-adjusted discount rate; if Mr Phillips is prepared to take a risk, the expected return of 8% is much better than the 3% return available for risk-free investments.

So, if Mr Phillips had to choose between undertaking the development this year or abandoning it irrevocably (which would have a zero return), he should definitely go ahead.

But this is not in fact the choice open to him. The permits aren't about to expire, and if he doesn't do the work this year there's no need to abandon it: he could do it next year, or the year after. He doesn't even need to decide whether he will do it next year or the year after – by doing nothing he retains the option to develop.

Mr Phillips needs to compare the value of developing this year, which he has worked out using standard whole-life costing, with the value of retaining the option to develop. His gut feel tells him that retaining the option to develop is an attractive proposition, but the numbers are missing.

The fringe of downtown Toronto, with carparks which really want to be office blocks. The sites will be developed as soon as the property market justifies investment – but no sooner.

The option to develop is one of the most common and important options in real estate decision-making. It accounts for many otherwise perplexing phenomena in the built environment. City centres are often surrounded by areas of what seems to be under-utilised land. It is hard for planners to predict when and how developers will proceed. Land is often held, apparently for development, for decades. Opportunities are not seized. Developers seem to demand unbelievably high levels of return before they proceed, so seem to be exceedingly greedy. With the addition of a few zeros, the lifecycle option for the development of Mr Phillips' basement or a city centre parking lot have a great deal in common.

Yet, it can be logical to pause, when rushing ahead with a project could lead to a waste of resources – the risk of over-investment. Surprisingly, sometimes procrastination is the right answer.

Mr Phillips may go with his gut feel approach to decision-making. However, we will try to model and quantify the option.

Responding to uncertainty

The main source of uncertainty faced by Mr Phillips concerns the future rental income from the new basement unit. After 20 years experience in the rental market, Mr Phillips knows about the risks. It is never possible to be certain of future rental income: rents will go up or down depending on the state of the economy, and supply and demand in the local housing market; in the worst case tenants might prove hard to find, leaving the unit empty for periods with no rental income. Based on his experience, Mr Phillips believes that rents could rise or fall by 10% per year. However, he does not expect any long-term trend for rents to rise or fall compared with the general economy, so the average value of his rent forecast would be just about the same as the current rent, in real terms. His forecast of the range of possible future rents can be represented on a binomial tree.

Mr Phillips' basement, presently an underused store. It is quite spacious, but radical transformation is needed to create a pleasant apartment.

Binomial tree of future rents

technique 6.2

The binomial tree starts with today's known rent, \$9,600, in Year 0.

In each successive year it is believed that rents could rise or fall by 10%. In Year 1 the higher rent and lower rents are given by:

$$\textit{higher rent: } v_1^u = v_0 \times u \qquad \textit{lower rent: } v_1^d = v_0 \times d$$

where u is the upward ratio, in this case (1 + 10%) = (1 + 0.1) = 1.1, and d is the downward ratio, here 1/(1 + 0.1) = 1/1.1 = 0.909. In this case $u \times d = 1$, but this need not always be true.

Note that it is a *recombining* binomial tree, since an upward move followed by a downward move gets to exactly the same value as a downward move followed by an upward move.

As well as the upward and downward ratios, the probabilities of a rise (p_u) or a fall (p_d) have to be specified. In all cases $p_u + p_d = 1$. The ratios are usually marked on the relevant branch, as in Technique Box 3.2. Here, all upward and downward probabilities are 50%, and as the tree is so small the numbers are not marked.

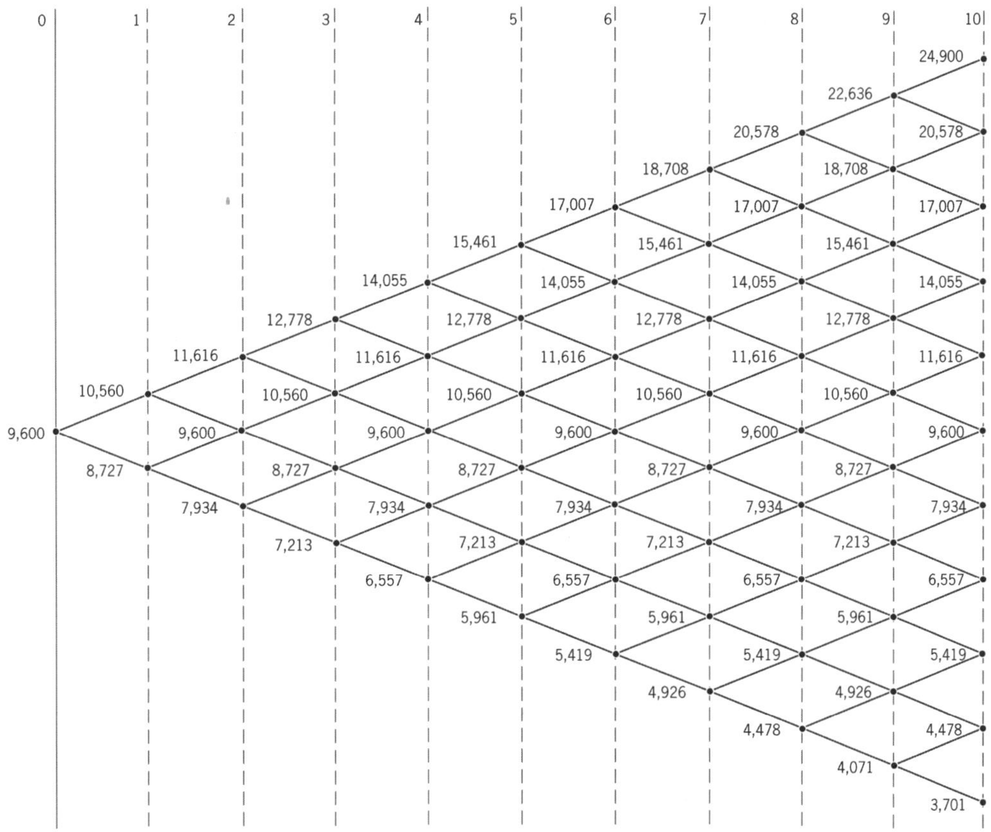

Although the upward and downward ratios are the same, the size of the upward steps is larger, so the average of the rents in each year rises gradually above the \$9,600 'median' line.

We will use the binomial tree to explore the value of alternative courses of action that Mr Phillips might follow.

Develop now

If the development is carried out now (Year 0), it will benefit from a future income stream starting next year (Year 1), discounted at 8%. The actual income at each year in the future is not known, but the range of possible incomes and their probabilities are included in the binomial tree. So the present value of the uncertain future income is the weighted average of every entry on the binomial tree, discounted back to Year 0. It comes to $65,860.

The net present value of the project is present value of the income stream less the development cost – it comes to $1,860. This is somewhat higher than the result of the standard WLC exercise, because the upward rent increments in the binomial tree are slightly larger than the downwards rent decrements (even though the upward and downward ratios are the same), but like the standard WLC has a positive value, indicating a viable project.

Wait a year

Suppose that Mr Phillips retained his option but it only lasted for a year – maybe those pesky permits were only valid for another 12 months.

In Year 1 the basement is still undeveloped and there is 50% chance that rents would have risen and 50% chance that they would have dropped. If they had risen, Mr Phillips would be faced by a different binomial tree of rents starting from a higher rental level. The present value of this stream of rents can be calculated, just as before. The present value is higher, $72,450, because the starting point was higher. The net present value of developing in Year 1 from the higher rental starting point would be $8,450.

On the other hand rents might have fallen. Again, there is a different binomial tree but this time with a lower starting point. The net present value of developing in Year 1 if rents had fallen would be –$4,130, not an attractive investment.

Present value from a binomial tree

technique 6.3

The present value of the entries in a binomial tree starts is the discounted weighted average of all entries. Let us look at the first three years only.

We build up another three of cumulative values. The calculation starts from the right hand side. The values for the last year are entered unchanged into the cumulative value tree. The entries for each node in the next year to the left depend on the nodes own rental income and the 'parent' nodes to the right to which it is connected by branches. The formula is:

$$V = v + p_u \times \frac{V_u}{(1 + r)} + p_d \times \frac{V_d}{(1 + r)}$$

where:

- V is the cumulative value at the node being studied
- v is the income at the node being studied
- V_u is the cumulative value at the upward parent node
- V_d is the cumulative value at the downward parent node
- p_u is the upward probability
- p_d is the downward probability
- r is the discount rate

For the highlighted node the calculation is:

$$V = 11{,}616 + 0.5 \times \frac{12{,}778}{1.08} + 0.5 \times \frac{10{,}560}{1.08} = 22{,}420$$

Starting from the right, this exercise is applied to every node until Year 0 is reached. The cumulative discounted value of all the entries on the tree of rents is the present value of the rental income.

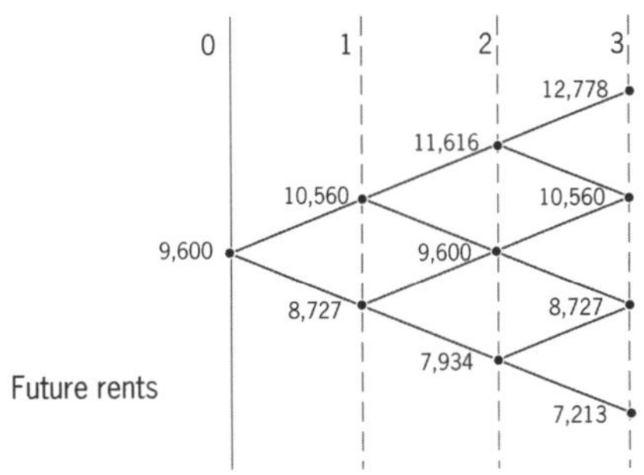

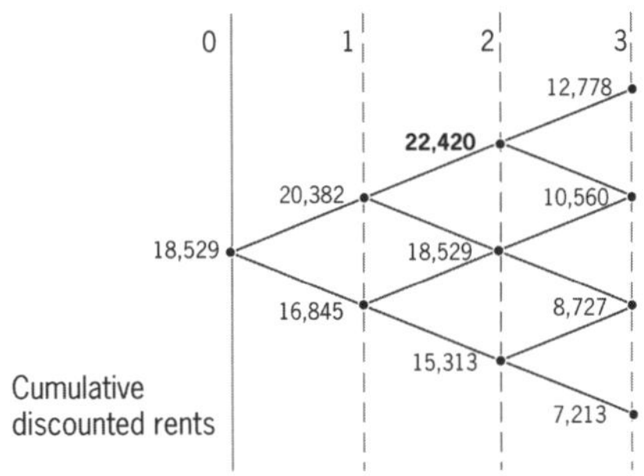

Using this method, the present value for the whole of Mr Phillips' binomial tree of rents, with zero income for Year 0, is $65,860.

If Mr Phillips waits until Year 1, there is 50% probability that he could undertake the development with an NPV of $8,447 and a 50% probability that he would abandon the project, yielding a value of zero (he wouldn't proceed with a negative NPV). If these two possibilities are discounted back to Year 0, the NPV of the weighted average of the two possibilities is $3,925. This is the NPV in Year 0 of retaining the one-year option to develop.

The NPV of retaining the option is higher than the NPV of developing in Year 0, which was $1,860: it is better to retain a one-year option than to develop. Mr Phillips' anxiety about pressing ahead was well-founded.

Perpetual option

In the one-year example, Mr Phillips has to decide after a year whether to proceed with the refurbishment or abandon it. If the option is longer lasting, at the end of the first year Mr Phillips can also choose whether to hold on to the option until the second year, and if he does he will then have the same choice in the third year, and so it goes on. If we allow the option to be held for 10 years its value increases compared to the one-year option. The value of retaining a 10-year option in Year 0 is $5,030, significantly higher than the NPV of developing in Year 0.

Option vs. development

This analysis of Mr Phillips' development opportunity is disturbing. It tells us that if there is an option to undertake a risky development, it can be better to retain the option than to proceed, even in situations where the NPV generated with a risk-adjusted discount rate says that it is a good investment.

This seems to violate the basic principle of net present value analysis, which says that projects with an NPV above zero are viable. Why?

It is not because of risk aversion, since that is taken into account in the risk-adjusted discount rate; it is simply that waiting ensures that you are able to make better decisions when better data is available maximising the likelihood of favourable outcomes.

Valuing the one-year option to develop

technique 6.4

If there is a one-year option to develop, and the option is retained (ie. no development is carried out in Year 0), then there are two equally probable outcomes in Year 1.

If the rent has risen between Year 0 and 1, the present value of developing in Year 1 is calculated in exactly the same way as for development in Year 0, but from a higher starting point. The first part of this binomial tree of rents is shown.

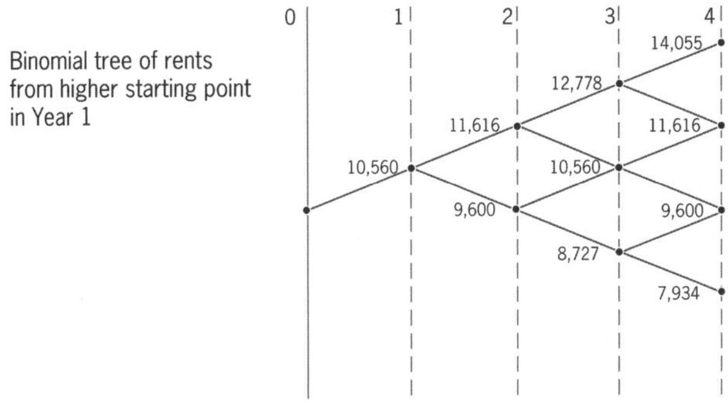

The present value in Year 1 of this binomial tree is $72,447. Construction cost is $64,000 so the net present value in Year 1 is $8,477.

If the rent has fallen between Year 0 and 1, the present value of developing in Year 1 is calculated as before, but from a lower starting point. The first part of this binomial tree of rents is shown.

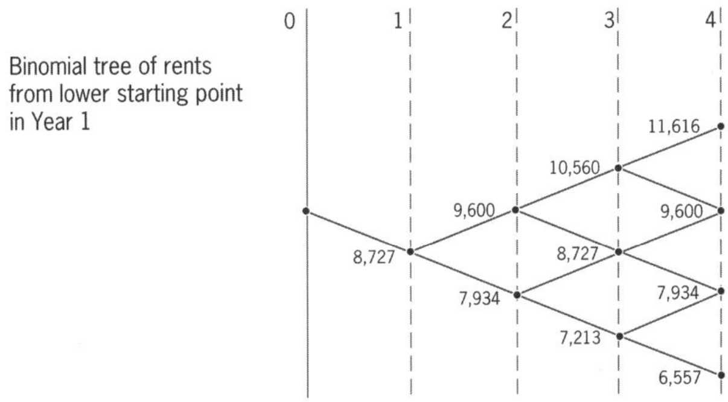

The present value in Year 1 of this binomial tree is $59,874 and the net present value after deducting the construction cost is –$4,126. With negative NPV this is not a viable project and it would be abandoned (remember in this version of the problem it is not possible to retain the option until Year 2), so it has zero development value.

Now move back to Year 0. The option to develop is retained in Year 0, and we now have the development values in Year 1 of the two equally probable alternatives. The weighted average of these alternatives, discounted to Year 0, is the value of retaining the option in Year 0.

The estimation of option value at Year 1 is

$$Q = 0.5 \times \frac{8,447}{1.08} + 0.5 \times \frac{0}{1.08} = 3,925$$

Valuing the 10-year option to develop

technique 6.5

Each node of the binomial tree of rents has a corresponding value for NPV if the project is developed with that starting rent. A portion of this tree is shown below:

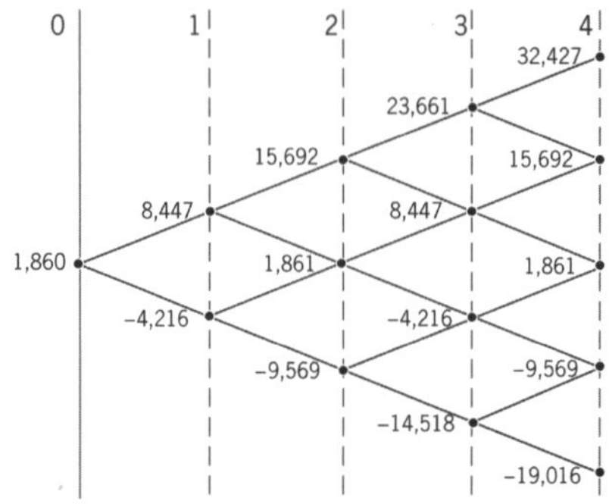

Some of the NPVs are negative, in which case the project would not be undertaken, yielding a zero value. We can construct a binomial tree of development values in which the negative NPVs are replaced with zeros:

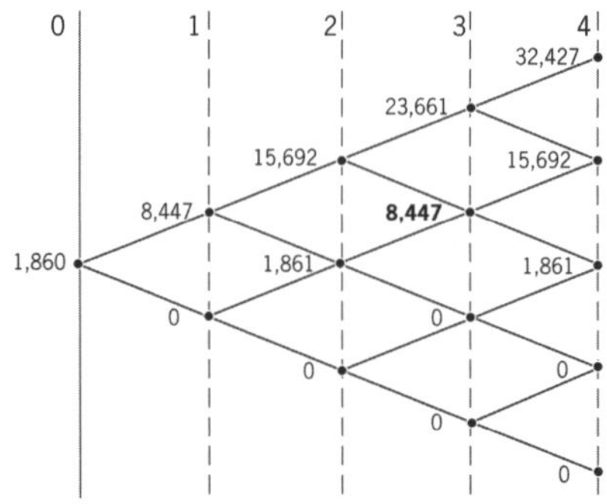

To get the option value for this tree we work from the right-hand side and carry all the discounted weighted average values back to Year 0, as before, but there is a difference. At each node we have to decide between two alternatives: develop, with the indicated development value, or retain the option in which case the development value at the node is the discounted weighted average of the two parent nodes. We cannot do both, so we choose the alternative with the highest value.

A part of the option value tree is shown below

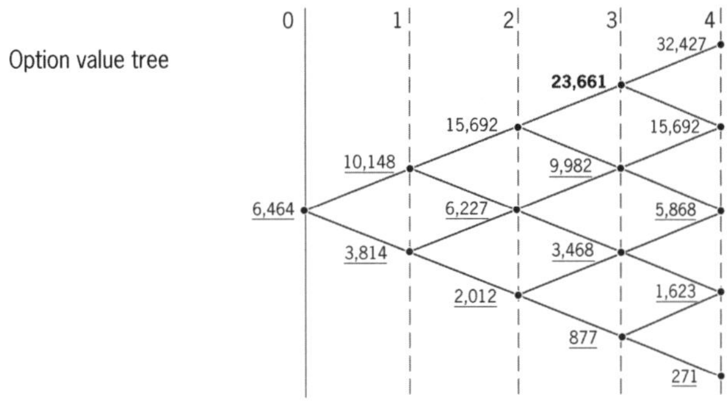

For the highlighted node, the development value is \$23,661, as marked in the development value tree. The value if the option is retained is derived from the two 'parent' nodes

$$0.5 \times \frac{32{,}427}{1.08} + 0.5 \times \frac{15{,}692}{1.08} = 22{,}277$$

At this node, development value exceeds option value and the project would be developed. At the node below, the development value is \$8,447 whereas the option value is \$9,982, so the option is retained. On this part of the graph, the nodes where the option is retained are underlined. Note that the project is developed only when rents have risen to a high level, otherwise the option is retained.

When the whole tree is brought back to Year 0 its value is \$6,464: this is the value of retaining the 10 year option at Year 0.

It can be seen that even when a project is 'far out of the money' the option still retains some value.

Does this mean that it is never a good idea to proceed with a development when there is an option to develop later? No – it depends on the numbers. If the NPV of proceeding now is close to zero, it is probably better to retain the option. But if the NPV of developing climbs well above zero, it will reach a point where the NPV of proceeding is equal to the NPV of waiting, at which point the project is viable.

In Mr Phillips example, a \$64,000 development with a starting rent of \$9,600 per year passed the NPV test but failed the 'option test'. If the development could be done for \$53,000 it would pass the 'option test' as well. Alternatively, if a \$64,000 development could command a starting rent of \$11,600 it would also pass the 'option test' – as would other combinations of construction cost and rent. These would be tough targets to meet – near 15% saving on the construction cost or 20% increase in rent. The option to develop has a dramatic impact on Mr Phillips project evaluation.

One could ask what rate of return would pass the 'option test'. For Mr Phillips' example, it turns out that the internal rate of return for the enhanced projects would be about 11.5%. This is higher than the risk-adjusted rate that Mr Phillips had set at 8%, representing an additional 'option premium' of 3.5%.

The message is this: in situations of uncertainty where there is an option to defer development, one should only proceed if the project has an exceptionally favourable return.

Meanwhile in the real world ...

These observations, which we have explored through options, are part of the wisdom of experienced managers. In risky situations where they have an option to develop, they don't go ahead with projects that have break-even NPV at the risk-adjusted discount rate. Observe the undeveloped parking lots in city centres where 'greedy' developers do nothing for years.

However, experienced managers have not been trained in 'options thinking'. To reconcile their gut feel with the NPV rule they crank up the risk premium in order to deter inefficient investment. This makes some sense, except that it is difficult to know how much to increase the risk premium; in practice it is usually a matter of guesswork.

The resulting cranked-up discount rate is called the *hurdle* for the project. It is higher than the risk-adjusted discount rate. The project cashflow – development costs and expected incomes – can be manipulated until a solution is found that clears the hurdle.

Needless to say, these numbers are specific to the data for Mr Phillips' example, and new calculations will be needed for other cases.

Interpreting the binomial tree technique 6.6

Suppose that the outcome of a project could be 'delight' or 'regret', depending on the future behaviour of an uncertain factor – rental levels in Mr Phillips' example.

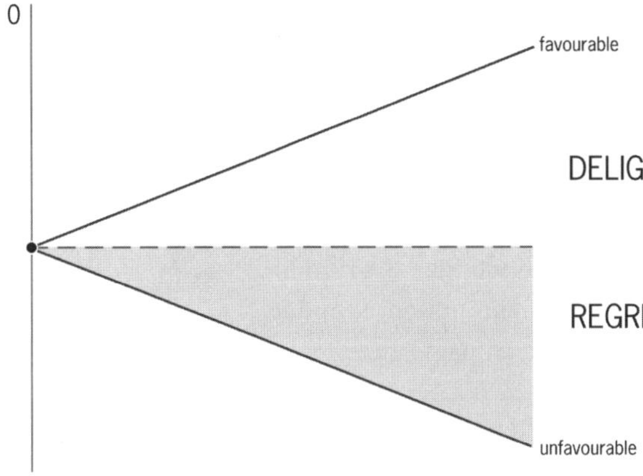

If this factor is just satisfactory in Year 0, as would be the case if the project has an NPV which is zero or just positive, then there is a good chance that the completed project will move into the 'regret' zone. How can it be avoided?

Rather than undertake the project in Year 0, an option allows the developer to wait and see how the uncertain factor performs. If it moves in a favourable direction, the project can be undertaken with a much lower probability of moving into the 'regret' zone. If it moves into 'regret' territory the developer will not go ahead, and be thankful that he didn't in Year 0.

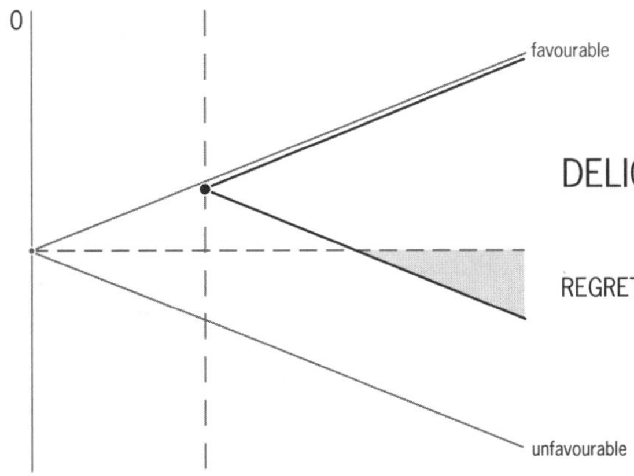

The essence of an options-based strategy is that the investor is able to make better decisions by waiting, because it is then possible to take account of information that is not available in Year 0.

Lessons from the option to develop

The option to develop is valuable because waiting allows one to benefit from better information about uncertain future events. The value of the option to develop depends on:

- uncertainty in the project outcomes: the greater the uncertainty the more the option is worth (if the future is certain there is no advantage in holding an option)
- the duration of the option: the longer the option lasts the more it is worth (if the option is about to expire, develop at the risk-adjusted rate).

These factors explain why property that is fit for development can remain undeveloped for a surprisingly long time, and is only developed when high returns are achievable.

The basic decision rules are:

- when there is no uncertainty in the project cashflow (a highly unlikely occurrence), invest if there is positive NPV at the risk-free rate
- when there is uncertainty but no option to defer development, invest if there is positive NPV at the risk-adjusted rate
- when there is uncertainty and an option to defer development, retain the option unless the NPV of proceeding exceeds the NPV of waiting.

If you do not follow these rules, there is the risk of under- or over-investment. Setting too high a hurdle risks under-investment; proceeding when it would have been better to retain the option risks over-investment. Both under- and over-investment are wasteful of resources and therefore damaging to sustainability.

The conventional NPV decision rule, which is the foundation of standard whole-life costing, produces incorrect decisions when there is uncertainty and an option to develop.

Centre Point (above) was developed in London in the 1960s and caused public outrage by remaining untenanted for many years. The developer was Harry Hyams:

Harry Hyams is a past master at the art of keeping buildings empty. This is not the crazy tactic it sounds. In a market where rents are rising it can pay hand over fist to leave a building empty almost deliberately.

Keeping a building empty is just a question of how much money you've got.

Oliver Marriott *The Property Boom* (1967)

The behaviour of experienced practitioners in the real estate and construction world is in accordance with these principles, but most practitioners are unaware of options and are often unable to explain how they arrive at their judgements, sometimes offering far-fetched explanations. The idea of the option to develop is conceptually simple and mathematically tractable: it is an extremely important idea for the rational evaluation of investment in real estate and property.

7

THE OPTION TO EXPAND

Many buildings are capable of expansion, so their owners have an option to expand. If expansion is easier to achieve, the option is more valuable. How can the value of the option to expand be assessed? The question is considered in this chapter, with the example of a residential project.

A quandary in the suburbs

In the spring of 2005 Drs John and Virginia Doherty, husband and wife physicians, were reflecting on their house, located in a pleasant suburb of a city in Tennessee. Since they purchased the house 20 years ago their two sons had completed university and moved away. In some respects the house seemed empty without the children, and the childrens' friends, cars and mess. However, it seemed that they were being visited by an increasing number of old friends, now similarly child-free, who were staying a few days. Yet again, they found themselves contemplating the space over the garage. A few years ago it seemed like a good idea to turn this storage space into a study, to allow the parents to escape the noise generated by the children. Now there is peace and quiet, and the Dohertys are wondering about creating a guest suite.

In principle, it should be easy to expand the house. It was built in the mid-1970s by a speculative builder to a somewhat standard plan, using a wood frame with brick veneer. Four tall columns across the front evoke images of the old south. Over the spacious two-car garage (a side wing attached to the house) is a large roof space, uninterrupted by trusses and accessed by a full size stairway. The space has gradually filled up with the detritus of 20 years of family life. By removing the junk and doing a bit of work, it should be easy to gain some useful extra space.

The Dohertys' developer house of the 1970s with an Antebellum portico but no headroom in the garage roofspace.

The Dohertys discussed this idea with some contractors, but were dismayed by their comments and estimates. The biggest problem was that the roof had been built several inches too low to make a habitable room, and it would have to be removed and replaced. The Dohertys couldn't help wondering why the original builder did not build the side wing roof a bit higher to allow for expansion.

The option to expand

It is clear that the space over the garage represents an option to expand the living area of the house. The option has existed since the house was built 30 years ago. However, it has a high exercise cost, as the roof structure has to be rebuilt. This option is one of a multitude associated with the house – a ground floor extension would be possible given the large site; perhaps a separate guest pavilion in the garden; a new garage could be built and the present garage turned into the guest suite; the whole house could be demolished and a new one built; and the Dohertys have the option to sell the house and move somewhere else.

Holding options is one of the benefits of property ownership, along with the enjoyment of the property in its current state. Although it is never possible to exercise every option, they all contribute to the value of the property.

With the low roofline over the garage the Dohertys have an option to expand the living space over the garage at a high cost; if the roofline had been built with more headroom they would have the option to expand at lower cost, but this would have required the outlay of additional resources when the house was built.

The interesting question is whether the original developer should have committed extra resources and built a higher roof, to create the more valuable option to expand – more valuable because it could be exercised at lower cost.

The pragmatic view

We don't know whether the developer intended to provide an option to expand in the roofspace over the garage. We can be fairly sure he wouldn't have called it an option, but he may well have been aware of the expansion potential. Probably no real analysis was undertaken, but the fact that the builder provided a stair suggests that he thought the space over the garage had some value.

The roofspace over the Dohertys' garage. Almost, but not quite, usable as an extension to the living space.

Quite possibly the developer, relying on experienced gut feel, decided that potential purchasers would not value the benefit of developing the roofspace enough to warrant even the small additional cost of making it easy to utilise. He knew that most purchasers' budgets are stretched to the maximum when buying a new house, especially if they have young children. They do not have spare cash to invest for uncertain and distant benefits.

Most young families expect to get wealthier over the years, so there is no sense in making present-day sacrifices when money is short, in order to receive benefits when there will be more cash available: immediate needs are given priority. Thus the expectation of increasing wealth leads to a short-term planning horizon.

There may have been other considerations – perhaps aesthetics? Speculative houses are often sold before construction so the perspective view is important. Maybe the author of the 'artist's impression' yanked the garage roofline down to make a better visual balance with his Antebellum portico?

The cost of expanding

There seem to be plausible reasons why the builder did not provide the option to expand at low cost, but let us put down some numbers. We'll work with real prices, so the effect of inflation is stripped out. This means we treat 1975 and 2005 building costs as being comparable, and for convenience we will use 2005 prices.

The Dohertys' house has a floor area of $200m^2$ and cost $\$1,500/m^2$ to build, giving a construction cost of \$300,000. The land cost was \$100,000 and the developer's profit

20%, so the sale price for the brand-new house was $480,000 (it was $45,000 in 1975 dollars).

The expansion over the garage would give 30m² extra floor area. If it could be constructed within the existing roofspace it would cost $1000/m² or $30,000 total cost; but if the roof has to be rebuilt it would cost $1500/m² or $45,000 total cost. Thus, if the original roof had been higher the saving in construction cost for the expansion would be $15,000.

If the original house had been built with a higher garage roof it would have cost $3,000 extra (at 2005 prices), one fifth of the cost of doing it as a retrofit. After adding profit, the developer would have charged $3,600 extra for the newly built house with the higher roof. The extra cost of $3,600 seems very small in relation to the $480,000 selling price, but you can be sure that there were numerous other extras that the developer could have added to improve the house – such as a high-style kitchen, extra bathroom, open fireplace, whole-house air-conditioning, and many others – and if he had done them all it could easily have turned into a $540,000 or $600,000 house and been out of step with the market. So every non-essential feature, including the option to expand within the garage roofspace, had to be justified.

The option to expand is one of the headline attractions of a suburban house. The option to build an extension would exist even without the planning permission, but permission greatly increases the option value.

The developer's test for any extra feature is whether it would be sufficiently valuable to prospective purchasers that they would pay the additional cost of providing it (or preferably more). So it is necessary to assess how a new home-owner would value the option to expand within the garage roofspace. If the home-owners valued this option at $3,600 or more they would prefer to pay extra for the house with the option; otherwise they would prefer the house without the option.

The value of the option to expand

How might one estimate the value of the option to expand? One could approach it in various ways, and the method we use below is not a standard technique but seems suitable for this particular case.

The trigger point for exercising this option is not a market indicator, but the occupants' desire for living space, as well as their level of wealth (too little and they couldn't afford the extension, too much and they would move to a larger house). In the example, the expansion had not been undertaken up to 2005.

It would be impossible for a new home-owner in 1975 to be certain whether the option to expand over the garage would ever be exercised, but maybe a probability could be estimated. Suppose that there is a 1 in 40 chance of expansion being carried out in any year – the first year after the house is finished, or the second year, or the third year and so on. If the house had been built with the higher roof, the owner would benefit from the $15,000 saving whenever the expansion is carried out.

A more sophisticated model might vary the probability of development over time, starting with a very low probability in the first few years after the house is built. Note that when the probabilities of expansion for each year are summed, there is no reason that they should add up to 1 – it is quite possible that the expansion is never carried out.

Going back to the simple model, in any one year the owner has a 1 in 40 likelihood of carrying out the expansion and saving $15,000, and a 39 in 40 chance of doing nothing to the roofspace. If nothing is done this year, the owner retains the option to do the project in the following years.

The option brings the possibility of saving $15,000 at various points in the future. The present value of this saving gives the value of the option to expand.

The present value of the possible saving depends on the home-owner's attitude to the future. A home-owner who attaches a high value to future costs and benefits will value the option more highly than one who is only concerned with the short term. The difference in attitudes to the future is expressed by the risk-adjusted discount rate: a long-term view corresponds to a low value for the risk premium and therefore a low discount rate, whereas a short-term view corresponds to a high value for the risk premium and high discount rate.

How to expand a house with a large Ionic portico? – add a small Ionic portico! Trelissick House, Cornwall.

Valuing the option to expand

technique 7.1

A tree representing the option to expand is built up from the left-hand side. In Year 0 the option to expand cannot be exercised, but in Year 1 there are two possibilities: either the option is exercised and a saving of \$15,000 (15k) is made, or the option is retained. If it is retained, in Year 2 there are two possibilities: either the option is exercised and a saving of \$15,000 (15k) is made, or the option is retained. Similarly in Years 3 and 4 and so on. The option expires in Year 40.

The 'upward' probability (p_u) at each branch is 1/40 = 0.025; the 'downward' probability (p_d) is 39/40 = 0.975. In all cases $p_u + p_d = 1$.

In this example the discount rate (r) is 8% per year.

The option tree is then valued, starting from the right-hand side. In Year 40 the value of implementing the option is 15k and of not implementing it is 0 (it cannot be retained). In Year 39 the value of implementing the option is 15k and of retaining it is the discounted weighted average of the two possible 'parent' outcomes, using the formula:

$$v = p_u \times \frac{v_u}{(1 + r)} + p_d \times \frac{v_d}{(1 + r)}$$

where:

- v is the value of holding the option
- v_u is the value of the upward parent node
- v_d is the value of the downward parent node
- p_u is the upward probability
- p_d is the downward probability
- r is the discount rate.

In Year 39 the value of retaining the option is

$$v_{39} = 0.025 \times \frac{15\text{k}}{1.08} + 0.975 \times \frac{0}{1.08} = 347$$

The same calculation is carried out for each year moving back through the tree to Year 0, when:

$$v_0 = 0.025 \times \frac{15\text{k}}{1.08} + 0.975 \times \frac{3{,}505}{1.08} = 3{,}512$$

The value of the option to expand at low cost is \$3,512, under these assumptions.

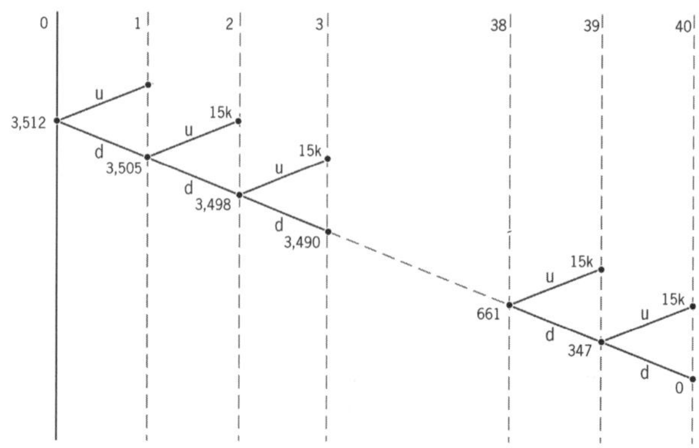

With a low risk-adjusted discount rate, the present value of the possible $15,000 saving will be high; whereas with a higher discount rate the present value will be lower. The table sets out a range of values for the discount rate and present value.

Discount rate	*Present value of savings*
3% (risk-free rate)	$6,059
5%	$4,742
8%	$3,512
12%	$2,576
20%	$1,666

Present value of possible savings from the higher garage roof, derived with different discount rates

The cost of creating the option would increase the purchase price by $3,600. The option would only appeal to purchasers with a long-term view, reflected by a risk-adjusted discount rate of 8% or lower, or to people who had a higher expectation of expanding the house. Remember that we are using real rates which are lower than market rates (because market rates anticipate inflationary growth).

What discount rate would be expected for a house purchaser? Mortgage rates are only just over the risk-free rate, because mortgage lenders protect themselves from risk so effectively: being much more exposed to risk, the householder would certainly have a higher discount rate. How much higher would have to be inferred from the individual household's consumption and saving behaviour: very few, if any, people could put a number to their personal discount rate for the option to expand.

Of course it was not the purchaser who made the decision on how high to build the roof, but the developer, who used his gut feel or expert judgement about the housing market of the 1970s. How did he do? We have slightly turned the question round to ask, 'did he make a good estimate of the home-owner's discount rate?' instead of, 'did he build the roof at the right level?' It seems like a close call. The Dohertys now wish he had made a different decision, but developed countries are now much wealthier than they were a generation ago, and assets that are affordable now were out-of-reach luxuries then. Maybe the Dohertys' option to expand fell into this category.

As in many cases, it is much easier to establish the cost of providing the option than to make a reliable estimate of the option value.

Prevalence of the option to expand

The option to expand is widely recognised and is a factor in many construction and real estate decisions. Practically every multi-phase masterplan ever proposed includes arrows pointing out from Stage 1 and marked 'expansion' or 'growth'. The arrows indicate an option to expand. The attraction of such an option is self-evident, but there are usually costs attached: land has to be kept vacant or in a short-term use that can be displaced. These costs have to be weighed against the value of the option, which like all options may never be exercised.

One might think that multi-storey apartments have no option to expand, but in Hong Kong cantilevered extensions create spectacular 'illegal facades'.

Many, possibly most, masterplans are never completed in accordance with the original intentions: any project that requires decades for completion is almost certain to be affected by unforeseen and disruptive events. Since this risk of non-completion should be anticipated, people who develop masterplans would be well-advised to ensure that each phase of development forms a satisfactory whole if work stopped. This is another option: the option to abandon the masterplan. It is perhaps just as likely to be exercised as the option to expand.

Some types of development have an embedded option to expand. Low density detached or terraced houses can be expanded into the garden or the roofspace, but in apartments the option is more restricted or non-existent. The option to expand almost certainly played no part when millions of utilitarian low density houses were built in Britain in the nineteenth and twentieth centuries, but the option was there and it has been widely exercised by the affluent post-World War II generations. The current pressure for high density housing in Britain will probably lead to fewer options to expand, perhaps reducing long-term value to society, especially if wealth continues to grow.

The El Gallo neighbourhood in Cuidad Guyana in Venezuela was laid out in 1964 with wide streets and generous plots, on the principle of 'progressive development'. This created the option to expand, and after 40 years the initial self-build shelters have been transformed into middle-class houses.

The option to expand was a major consideration in a self-build housing scheme in Venezuela of the 1960s. Rather than allowing tightly packed structures, which are typical of self-build settlements, the site was laid out with wide streets and large plots, even though the initial dwellings were small and flimsy. Over a generation the initial dwellings were upgraded and expanded by their owners to create a thriving housing area. A tighter layout would have become congested, inhibiting the possibility of improvement. The option to expand, deliberately created by the initial low density, facilitated the development of the community.

These examples of the option to expand have been concerned with the physical expansion of buildings. Similar processes apply to space within buildings. An organisation renting, say, 10,000ft^2 of office space in an office block might have a possible but uncertain need to expand by another 6,000ft^2. Renting 16,000ft^2 of floorspace would clearly secure an option to expand, but very expensively. Some commercial landlords cater for this situation by offering leases with an option to expand: if the trigger point is reached, the tenant can exercise the option and require the landlord to provide more floorspace. The rent is increased for a lease with this option, but it is much cheaper than renting extra floorspace. A landlord who offers the option to expand cannot afford to leave space empty, so must arrange short-term tenancies which can be terminated if the main tenant's option is exercised. The landlord benefits from the option arrangement through the higher rent paid by the primary tenant (although the short-term tenants will pay lower rents), but more importantly by a reduction in the risk of the primary tenant relocating to another building to satisfy a need for growth. As well as the option to expand, leases can contain the option to contract, so that the tenant can relinquish unneeded floorspace, again subject to a rental premium. These lease options are very clear examples of the use of options to manage uncertainty.

Denys Lasdun's masterplan of the 1960s for the University of East Anglia was unfinished, but fortunately this had been anticipated:

A significant quality of Lasdun's design is that ... the project can be modified and elaborated without damage, and other architects ... can be called in to contribute to it as new requirements become apparent.

Architectural Review (1969)

Destroying the option to expand

It is important to be alert to the risk that the option to expand can be curtailed or destroyed. For example, the Botanic Garden in Cambridge acquired a 40 acre site in 1830 but only half was laid out; the other half was rented out as allotments until expansion was affordable. In the 1920s a director's house was built on some of the allotments, but it was placed in what proved to be a most awkward position when the Garden finally expanded to its full 40 acres in the 1950s: evidently the impact of the director's house on the option to expand was overlooked, or the option was assigned a minimal value.

Sometimes external forces intervene to destroy the option to expand. For example, a building that was scheduled for demolition to make way for expansion might become listed as a historic structure and therefore protected from demolition, as at the Fitzwilliam Museum in Cambridge. A similar difficulty has arisen at some of the British new universities of the 1960s: they had large sites to allow room for expansion, but early phases of development were modest so the buildings were surrounded by large open spaces. This has become so well-liked that some universities now have great difficulty in getting planning consent to expand into the land that was specifically acquired for expansion.

Problems can arise if development is started in the belief that there is an option to expand, but this turns out not to be the case. For example, the History Faculty building in Cambridge was designed by James Stirling in the 1960s in the expectation that a crucial parcel of land would come into the University's ownership, but when construction was due to begin the former owner was still in possession. There was no time to redesign the building so the plans were turned round. The building is permanently burdened by having its entrance at the back and excessive south-facing glazing.

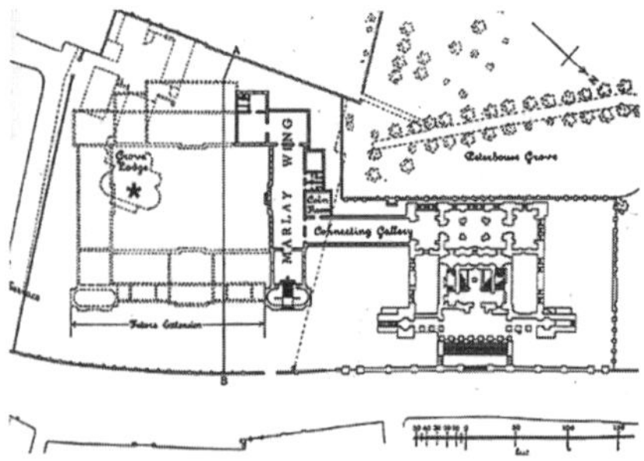

An ambitious expansion of the Fitzwilliam Museum in Cambridge, planned in 1914 (above), involved demolishing an elegant villa of the 1790s which is now listed (marked with an asterisk). A crisis was averted because funds ran out in the 1950s. The 'temporary' end wall of the extension is now a permanent fixture – new extensions have been built in front of it (below).

Summary of the option to expand

Although the option to expand is prevalent, we expect that it is rarely quantified and often under- or over-valued. Under-valuing the option means that future expansion is made unnecessarily expensive or impossible because an available option was not acquired or retained. Over-valuing the option is when resources, which could be used for more pressing and productive uses, are tied up to allow for some vague or improbable expansion possibility.

The options framework provides a way of clarifying the issues and prioritising the option to expand in relation to other claims on resources.

Where appropriate, it is a very good exercise to put numbers to both the cost of acquiring or holding the option to expand, and its estimated value. Opportunity costs are vital in such a comparison: vacant land may appear to provide a no-cost option to expand, but the value of alternative uses of the land is an opportunity cost of keeping it vacant.

The cost of an option to expand is usually easier to estimate than its value. Like other options, the value of an option to expand depends on:

- *amount of uncertainty*: options are more valuable when there is more uncertainty
- *duration of the option*: long-lasting or perpetual options are more valuable
- *trigger point*: the option is more valuable if the trigger point at which it would be exercised is more likely to be reached
- *cost of exercise*: the option has greater value if there is a low cost attached to exercising it
- *resulting benefit*: the option is more valuable if there are large benefits when it is exercised.

As always, the value and cost of the option to expand have to be compared, to assess whether it should be acquired or retained.

8

THE OPTION TO SWITCH USE

When a developer is faced by evenly-matched alternatives, a flexible solution that works for both alternatives is attractive. But how much is it worth paying for flexibility which may never be used? The question is explored in a city centre development: should it incorporate a switch use option for office or retail use on the ground floor. This option is quantified using a simulation model. Other examples of the option to switch use are discussed.

Venturing into a new market

Connfield Projects Ltd is a medium-sized development company which has successfully developed many suburban office buildings, serving the needs of innumerable small and medium-sized companies. However, Peter Holden, the company's founder, president and majority shareholder, has just acquired a site in the centre of a major city. His staff is surprised that he has made such a radical change in direction. However, Mr Holden feels that this diversification represents an opportunity for the future growth of the firm. The office building that can be accommodated on the site will give his firm much-needed profile as more than just a developer of low-rise commodity buildings in the suburbs.

As an alternative to his usual architect, Mr Holden has retained Howard Herzberg, who has substantial experience in more prestigious urban structures. However, the Connfield staff has noticed that their employer and the new architect are not always in complete agreement. One issue has arisen with regard to the ground floor of the building. Mr Holden has had little experience with retail and has in the past seen it as a nuisance, while Mr Herzberg has pointed out the possible value of ground floor retail space in a city centre location. Mr Holden is not convinced that retail space makes much sense, but

Ground floor retail units in a new city centre office building. If they remain unlet the landlord may consider switching to office use of the ground floor.

Commercial rent processes

principle 8.1

Commercial rents are set in a responsive but complex market. Unlike financial transactions which are highly standardised, every commercial lease is site-specific and tenant-specific. Retail rents vary substantially depending on the exact street frontage and layout of each space. Matters are confused by the inducements that are offered to tenants, including rent-free periods. Landlords prefer to offer rent-free periods rather than lower the rent, as the capital value of a commercial building is usually based on a multiplier of its rent.

Data from past transactions gives an indication of the general level of supply and demand in a particular locality, which is likely to correlate with rising or falling rents in a specific building, even though the rent for the building may differ considerably from the market average.

An effective way of simulating rents is to use a binomial tree in which rents rise or fall in each time increment (year, quarter, month, week ...). Any path through the binomial tree is a possible scenario of future rents – there are very many such paths, reflecting uncertainty about future rents. For the case study, data compiled from a variety of sources of the usual dubious reliability was used to develop reasonable binomial trees for office and retail rents. Market rents were generated twice per year. The tree had no upward or downward trend: long-term real growth or decline is very difficult to sustain because the market responds through the laws of supply and demand. If rents rise new space is created, in due course curtailing rent increases – a mean-reverting process. A specimen path through the binomial tree of rents is shown below.

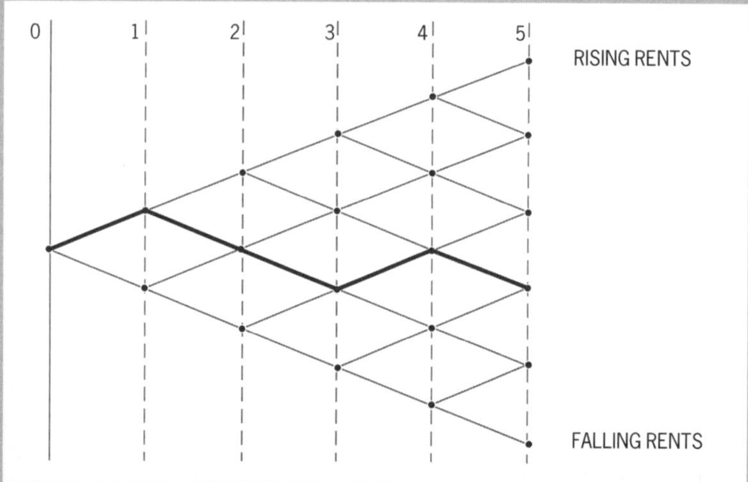

The property market varies on quite a short-term basis, but when commercial property is leased the rent is fixed for the duration of the lease or until a rent review. In the case study, five-year periods are assumed. The rent for the period is set at the current rent on commencement or renewal, and remains fixed – unless the tenants default, which would lead to a new lease at a new rent.

In the case study in this chapter, we have simplified the binomial tree and shown five-year increments. In a computer simulation it makes sense to work at the level of precision of available data, using annual or monthly increments: a coarser model does not make best use of available data, but an excessively precise model would be a wasted effort.

he has asked Connfield's in-house MBA, Ms Amber Guennel, to assemble some data.

The ground floor use dilemma

Connfield's problem results from a lack of familiarity with city centre retail space, which means that Mr Holden's gut feel is not as effective as it would be in more familiar situations. His inclination is to avoid retail. Mr Herzberg, on the other hand, has dealt with many similar developments, and is willing to pass on his insights.

In some city office buildings it would be very difficult or impossible to switch the ground floor use to retail (above), but in other cases the option is much more easily exercised (below).

Ms Guennel's enquiries quickly showed that, at the present time, the rental level for retail space at the location of the new building is lower than for ground floor offices. However, rents are volatile and the character of the area is changing, so in the future there is a definite possibility that ground floor retail space could command higher rents than offices.

When the relative benefits of ground floor retail and office space are uncertain, it is attractive to think of a flexible design that could start off as office space and change to retail use if a shift in rents occurred. A flexible design would be more expensive to build, with grade-level access, glazing and doors suitable for retail tenants, scope for retail tenants' security, deliveries, sanitary provision and means of escape, and perhaps a separate mechanical system for the ground floor.

Spending some money now would allow Connfield to switch the space to a retail use in the future – it would create the option to switch use. Would it be a worthwhile investment?

The crux of the decision is one that occurs frequently in real estate and design – how much should one pay for flexibility? It is not a brilliant insight to state that flexibility is generally beneficial, but it is not usually clear how much it is worth paying to get it. Even many years after decisions have been made and outcomes known, it is not necessarily clear whether flexibility has justified its cost.

As is often the case with options, it is reasonably easy to estimate the cost of providing the option to switch use, but valuing the option is a greater challenge.

Valuing the option to switch use

The option to switch the use of the ground floor space between retail and office use is valuable if there is significant uncertainty about which use will command the higher rent in the future. The option is attractive because of uncertainty concerning future rent levels. To calculate option value it is necessary to quantify the uncertainty. This is not a prediction, which is impossible, but a probabilistic forecast which should be based on the best available data.

The ground floor use of a college building in Cambridge was switched to retail use from storage, to take advantage of the frontage onto a popular shopping street. Formerly the building had a blank wall to the street. The cost of exercising the option was high.

As usual, data presents a major problem. Connfield has no experience of its own upon which to draw. There are some property market reports giving office data, but retail information is very limited and tends to be anecdotal – it is harder to generalise when retail rents vary dramatically on a street-to-street basis. Ms Guennel is unlikely to find high quality retail data, especially time-series data from which reliable estimates for rental volatility over time could be made.

If Ms Guennel could assemble detailed time-series data for office and retail rents, it would be possible to carry out a sophisticated mathematical analysis and calculate a precise number for the value of the option to switch use. This is how financial analysts establish precise values for financial options. However, the poor quality of most real estate data means that the effort put into sophisticated analysis would be wasted. Furthermore, in the construction industry there are very few people with the kind of mathematical skills that are found in financial institutions. In the real estate world, sophisticated models are unlikely to yield worthwhile benefits. Simulation is a promising alternative.

Simulating office and retail rental income

For the office/retail switch use option, a simulation model can be quite uncomplicated. Ms Guennel decides to use two binomial trees to project possible future office and retail rents. The trees start with today's known rents, and then rise or fall in each time period of the simulation according to ratios that have to be defined. The larger the ratios, the greater the 'spread' of the binomial tree. This corresponds to the amount of uncertainty, or the volatility, of rents over time.

Historic data indicates that the binomial tree for office rents should have somewhat higher volatility than that for retail rents. A 50/50 probability of the rents rising or falling in each time period is assumed, because there is no reason to expect any long-term trend of rising or falling real rents.

An important question is whether the office and retail rents tend to go up and down independently or in step

Simulation

principle 8.2

Simulation is an effective way of exploring 'what-if' scenarios in an intuitively convincing way. One benefit of simulation is that it forces the modellers to reflect upon each piece of data and the underlying structure of the problem and cause-and-effect relationships. This is why it is preferable for a project team to be fully involved, rather than just assigning the simulation to an 'expert'. Simulation is not simply a matter of putting in the data, 'cranking the handle' and getting an answer.

Simulation is a good way of dealing with multiple sources of uncertainty, such as two or more rent streams, or perhaps varying construction or operating costs. The interaction of two different sources of uncertainty is extremely hard to calculate mathematically.

Decision-makers, who rarely have advanced mathematical knowledge, are often unsure whether complex mathematical calculations actually describe the situation of interest: some false assumption or minor glitch could easily make the results simply wrong. A simulation has the benefit of being more transparent.

Perhaps the greatest advantage of simulation is that it requires a degree of mathematical knowledge which many managers retain from their schooling.

Older sources suggest that the simulation model and all the input data should be finalised before the model is run: this was because of the high cost of data-processing. Now it is feasible to make multiple runs to fine-tune the model and data.

There are some problems that can arise with simulation:

- as simulation is a random process, there is a chance that the results may be untypical; this risk is reduced by increasing the number of runs of the model
- the simulation model may not, in fact, be an accurate representation of the underlying reality, which may not be well understood because of problems with poor data
- simulations have a way of becoming ever more complex, as additional variables and interrelations are included, compromising simplicity and ending with something the decision-makers may not fully understand.

Simulation of the rental scenario

technique 8.1

The simulation begins with two binomial trees setting out the range of possible future office and retail rents. Each binomial tree starts with the known rents, £185/m² for office use and £150/m² for retail use. Based on data for past volatility, the upward and downward ratios are 6.8% per half-year for the office tree and 5.4% per half-year for the retail tree; the upward and downward probabilities are 50% in both trees. The first few time periods for the two trees are shown below:

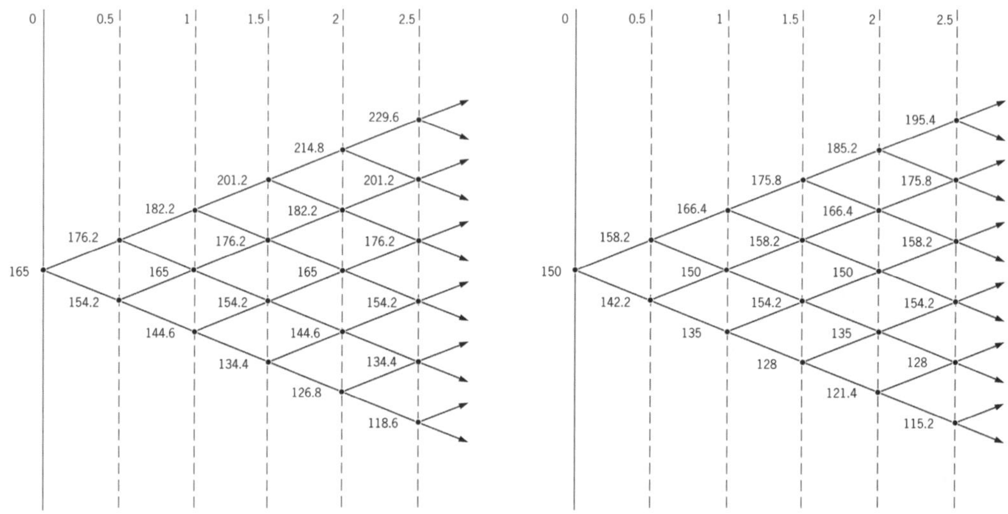

The simulation generates a 'path' through each of the trees, starting from the present rent and using a random number generator to select an upward or downward move in the next period. The two rental paths together constitute a single simulated rental scenario.

Trillions of different rental scenarios could be generated, because there are so many paths through the binomial trees.

The rental trees have steps every half-year, reflecting the volatility of commercial rents, even though the leases for the Connfield development have five-year duration. A typical rental scenario is tabulated below in terms of the rental levels for five-year leases.

Year	*1-5*	*6-10*	*11-15*	*16-20*	*21-25*	*26-30*	*31-35*
Office	£165.00	£188.20	£165.00	£144.60	£201.20	£229.50	£214.80
Retail	£150.00	£135.00	£158.20	£206.00	£254.40	£228.40	£167.00
Preferred use	OFFICE	OFFICE	OFFICE	Switch to RETAIL	RETAIL	RETAIL	Switch to OFFICE

The table also shows the preferred use for the lease periods, if there is an option to switch between office and retail. Note that in years 26-30 the office rent is marginally higher, but not sufficient to justify the cost of exercising the option to switch from retail to office use.

Correlation

principle 8.3

Two processes are correlated if they behave in similar ways – when one rises the other rises, and when one falls the other falls. Correlation is found when one process causes the other, like hours of work and number of mistakes, or if they are both caused by some other factor, like rising house prices and rising tender prices which are both caused by an economic boom. Sometimes observed correlation may be a coincidence, in which case the correlation is unlikely to continue for very long. If there is a reason for past correlation that remains valid, the correlation can be expected to continue.

Perfect correlation means that the two processes are locked together. If two processes are completely independent there is zero correlation. Between these extremes there are varying degrees of correlation.

There is inverse correlation between two processes if one rises when the other falls – a see-saw effect.

with each other, that is, whether or not they are correlated. The data collected by Ms Guennel suggests that the correlation is low – sometimes retail rents are rising while office rents are falling. In her simulation the two rental streams are treated as independent. This simplifies the model, but if there is correlation which the simulation model underestimates, then the option will be over-valued; this is because there is more uncertainty in uncorrelated rents and option value increases with uncertainty. If significant correlation exists it must be included in the simulation model.

For her model Ms Guennel assumes that rents will be fixed for five-year periods. If the building has an option to switch use, it could be exercised at the beginning of each of these five-year periods. When the option is exercised the building owner incurs the cost of making alterations, and may lose rent while the alteration work is in progress. This cost is estimated to be £250,000 each time the option is exercised.

Each time Ms Guennel's simulation model is run it generates one out of the vast number of possible rental scenarios. From each scenario the corresponding rental income can be calculated, with or without the option to switch use. This is the project cashflow.

The cashflow has to be discounted to find the present value of the rental income. What discount rate should be used? Commercial property development is a risky

Cashflow and present value for a rental scenario

technique 8.2

Using the typical rental scenario from Technique Box 8.1 we can calculate the rental income in each of the five-year periods: it is five times the annual rent/m² multiplied by the ground floor area, 1,000m². This is tabulated.

Office and retail incomes	*Years 1-5*	*Years 6-10*	*Years 11-15*	*Years 16-20*	*Years 21-25*	*Years 26-30*	*Years 31-35*
Office	£825,000	£941,000	£825,000	£723,000	£1,006,000	£1,147,500	£1,074,000
Retail	£750,000	£675,000	£791,000	£1,030,000	£1,272,000	£1,142,000	£835,000

There are now two cases to compare.

1. No option to switch use

If there is no option to switch use, the five-year cashflow of ground floor income is equal to the office rental data given above. The present value of the cashflow is calculated in the normal way using the approporiate discount rate for office rents, 8% per year (in real terms). For the NPV calculation the five-year cashflow shown above is expanded to give 35 entries, one for each year. The NPV for this rental scenario is £2.05m.

2. Option to switch use

In order to establish the cashflow it is necessary to determine whether the option is exercised. In Years 16-20 the retail rent is higher than the office rent, by £307,000. It costs £250,000 to alter the building, but the difference in rental income is sufficient to justify exercising the option. The cashflow for Year 16 is reduced by the cost of switching. The option is exercised again in reverse in Year 31. Note that in Years 26-30 the ground floor is in retail use even though office rental is slightly higher – but not sufficiently higher to justify exercising the option. The resulting five-year cashflow is shown below.

Switch use cashflow	*Years 1-5*	*Years 6-10*	*Years 11-15*	*Years 16-20*	*Years 21-25*	*Years 26-30*	*Years 31-35*
Use	OFFICE	OFFICE	OFFICE	RETAIL	RETAIL	RETAIL	OFFICE
Net income	£825,000	£941,000	£825,000	£1,030,000 – £250,000 = £780,000	£1,272,000	£1,142,000	£1,074,000 – £250,000 = £824,000

To get the NPV of the switch use cashflow, it is expanded to give 35 entries and discounted. The discount rate is lower than the 8% per year used for the office use cashflow, as there is less uncertainty and risk with the switch use option. The reduction of risk indicated by the simulation shows that the discount rate should be adjusted to 6.67% per year (in real terms). With this discount rate, the NPV of the switch use cashflow for this rental scenario is £2.41m.

The value of the option to switch use is the difference between the NPVs with and without the option. For this typical rental scenario, the value of the option is £2.41m – £2.05m = £360,000. For other scenarios there will be a different option value.

business, requiring heavy financial outlays with uncertain future returns. Mr Holden knows all about this – at least he does for suburban offices. For his first venture into the city centre the uncertainty is greater, and this is reflected in a higher risk-adjusted discount rate. He starts with a discount rate of 8% per year: with simulation models it is always possible to alter assumptions like the discount rate and re-run the model. Real prices and real discount rates are used.

Small-scale exercise of the option to switch use: a kiosk in the ground floor of an office block in the City of London.

Using the chosen discount rate, net present values are calculated for the simulated rental cashflows, with and without the option to switch use. The difference between the two NPVs gives the value of the option to switch use, for that particular rental scenario.

The present value of future rental income *with* the option to switch use is practically always equal to or higher than the present value *without* the option, because the option would never be exercised unless it generated extra income. Only freak circumstances could lead to a lower present value with the option, and with enough simulations these will occasionally appear.

When Ms Guennel's model is debugged and working, it is used in a 'Monte Carlo' simulation, when it is run hundreds or thousands of times. Each run generates a rent scenario and a corresponding value for the option to switch use; the results from each run are kept and analysed to indicate the average value, or expected value, of the option. This can be compared with the construction cost for creating the option, which has been estimated by Mr Herzberg. If the expected value of the option to switch use is greater than the cost, it is a good investment; otherwise it is better to build a cheaper building in which the ground floor is only intended for office use. Switching might not be absolutely impossible, but the cost would be prohibitive.

The first Monte Carlo simulation for the Connfield development suggests that the expected value of the option to switch use for the ground floor would be £424/m^2, which is higher than Mr Herzberg's estimate of £250/m^2 for the cost of creating the option; so at first sight the option appears to be a good investment.

Monte Carlo simulation

technique 8.3

Every time the simulation model is run it generates a new rental scenario, new values for NPV with and without the option to switch use, and the corresponding value of the option to switch use for that scenario (the difference between the two NPVs).

In a Monte Carlo simulation the model is run many times, and a record kept of the values of the option to switch use. Other data can also be recorded, for example the number of times the option is exercised. The data can then be analysed, to establish the average value of the option and the variance around the average. The average number of times the option is exercised and other statistics can also be generated.

With enough runs a Monte Carlo simulation can create a probability distribution of outcomes. One common mistake in simulation is not having enough sample runs. Fortunately, spreadsheets make it easy to undertake hundreds or thousands of simulation runs.

In this case the Monte Carlo simulation comprised 1000 runs with an average value of the option to switch use of £424,000. The distribution of values is shown below:

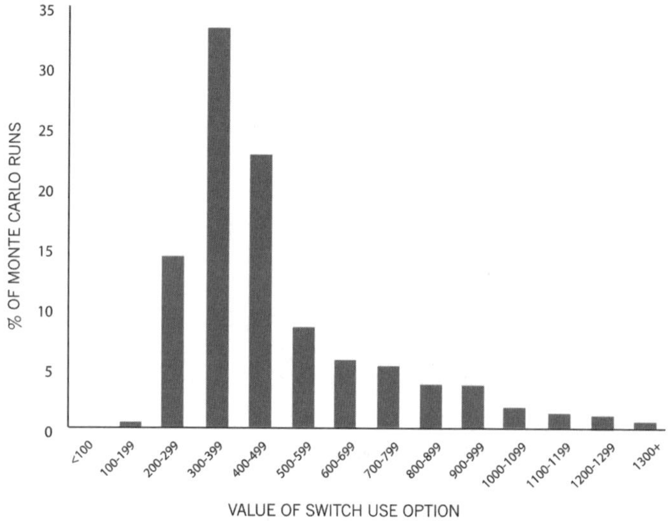

On the basis of this simulation data the project team will decide whether to invest in the option to switch use. But before finalising their decision, they may well come back and re-run the Monte Carlo simulation with different input data, for example amended values for the cost of exercising the option, the discount rate, or the upward and downward ratios in the binomial trees: this is sensitivity analysis. One of the strengths of simulation is that it is easy to make 'what-if' explorations.

Before finalising the decision the bionomial tree assumptions are double-checked and more thought given to the discount rate; alternative ways of designing the ground floor are assessed, with trade-offs between cost of acquiring the option and the cost of exercising it (these are usually inversely correlated). Using the simulation model as an integral part of design development, the project team decides that it is most cost-effective to adjust ground floor slab level and ceiling height to suit future retail use, and arrange the plan layout and mechanical services to ensure that ground floor retail premises could operate independently of upper floor offices, but not to provide dual-use window systems. The initial outlay is more modest than some alternatives, and the cost of exercising the option would be higher because the windows would have to be changed. However, with the data from the simulation, Peter Holden is confident that he has neither under-invested nor over-invested in the option, but pitched it about right, given today's uncertainty about future rents.

There are other switching options that could be studied in a similar way. One is the option to switch fuels: a dual-fuel boiler would allow the building owner to switch between oil or gas or even solid fuel, depending on uncertain future fuel prices. The extra cost of a dual fuel boiler buys the option to switch fuels; the value of the option could be estimated by simulation.

New uses for old buildings

Many buildings have an embedded option to switch use, as evidenced by the conversions that are now commonplace: warehouses become offices, offices become apartments, barns become cottages, etc. The names of many recent arts centres celebrate their non-arty past – the Brewery in Kendal, the Drill Hall in Lincoln, the Corn Exchange in Cambridge, and so on.

Some before-and-after combinations are more plausible than others. Multi-storey carparks, especially the warped slab sort, are thought to be one of the building types with the least opportunity for switching use.

Many switches of use benefit from what appears to be redundancy in old buildings – like the generous room dimensions and ceiling heights, and massive structures, found in nineteenth century warehouses. However, we can assume that flinty Victorian businessmen, who were willing to consign paupers to the workhouse in their own day, did not put redundant resources into their warehouse buildings so that they would make better loft apartments for City analysts 130 years later. The original developers must have built for the demands of their day with the technology of their day, and the lack of modern engineering materials and knowledge meant that the buildings were more robustly constructed than would be the case today.

Multi-storey parking garages are considered to be a building type with very limited potential for switching use.

New buildings as well as old are switched to new uses. An extreme case is the new local government office in Suffolk, in eastern England. This occupies a building in Ipswich that was designed as the regional headquarters of an electricity supply utility. Halfway through the construction period the utility company was taken over and the new headquarters was no longer needed. Suffolk County Council bought the half-finished building and completed it to meet their needs; in 2004 they moved into their new council offices.

Designing for the option to switch use

Many architects recognise that there is uncertainty about the future use of the buildings they design, and they seek design responses that facilitate future change. There is a danger that ingenious proposals may over-invest for benefits that are very unlikely to be realised.

Just as the inherited redundancy of old buildings facilitates new uses, redundancy is an important strategy for creating the option in new buildings. Enhancing the specification with features like, for example, larger spaces, taller ceilings, stronger floor slabs or larger service ducts, will undoubtedly enhance the option to switch use. However, redundancy usually has a cost penalty.

The notion of a universally adaptable building that can accommodate any conceivable demand is seductive. It is probably unachievable, and for low uncertainty situations it would be over-investment. Conversely, where there is very high uncertainty it is doubtful that an organisation would want a long-term commitment to any building, even an adaptable one. Designing adaptable buildings is not an abstract exercise, but has to be context-specific.

The idea of buildings characterised by 'long life, low energy, loose fit' was promoted in the 1970s. Simple, spacious and versatile Georgian buildings were taken as a model. The argument was that buildings which are optimised to a particular function at a particular time are likely to become obsolete when activities change – as they surely will. It was suggested that an alternative to optimisation is redundancy: to accommodate the same activities, a loose-fit building will be slightly larger than an optimised building, but it will also accommodate a wider range of alternative activities.

This insight was plausible but hard to quantify. Using the options approach, quantification becomes possible. The redundancy of loose-fit buildings can be seen as the cost of acquiring the option to switch use, and simulations such as the one in this chapter can provide estimates of the value of the option.

The loose-fit idea was applied by the architect John Weeks in the 1970s, in the hospitals he designed as 'multi-strategy buildings'. Now that these buildings have experienced a generation of activity change, Weeks' experiments – for that is what they were – could provide a wealth of empirical data. Did the buildings have too little or too much redundancy, or the right amount? That is, did they under- or over-invest in the option to switch use, or get it about right?

This discussion has assumed that a building acquires the option to switch use through paying for some degree of redundancy. Sometimes there is no cost penalty, when the option to switch use can be enhanced by improving the arrangement of spaces, services, etc, without adding to quantity or complexity. These opportunities should be identified and exploited.

The Corn Exchange in Cambridge was begun during an agricultural boom and finished in 1875, just in time for the next agricultural depression. After World War II it was reduced to marginal uses, but its spaciousness and central location led to conversion into a new concert hall, which opened in 1985. The switch use option from corn exchange to concert hall was inherent in the Victorian design, although it could not have been a conscious design objective.

Summary

The option to switch use applies in many different situations, in both existing buildings and new designs, and always adds value. How much value depends on context-specific factors which can be considered under five headings:

- *the amount of uncertainty*: the option will be more valuable if there is greater uncertainty about the future use of a building, for example when a building is for a new and risky enterprise
- *the duration of the option*: long-lasting options are more valuable, and the option to switch use is usually perpetual, being inherent in the configuration and specification of a building; sometimes legal or planning restrictions may curtail (or open up) the option, for example if there is change in the zoning of building uses
- *the trigger point*: the option is more valuable if it is more likely to be exercised; the option to make a highly improbable switch of uses will have low value
- *the cost of exercising the option*: the option is more valuable if the switch of uses can be made with little or no cost
- *the resulting benefit*: the option is more valuable if the switch of uses would generate large benefits.

By comparing its cost and value, the relevance of the option to switch use can be evaluated in particular cases.

The National Trust's visitor centre at Anglesey Abbey is due for replacement, and the NT wonder about running conferences in the replacement building. As this is a new and risky venture, it would be prudent to build the conference facilities with the option to switch use if the new business is not viable.

9

THE OPTION TO RECONFIGURE

Buildings are usually longer lasting than organisational needs. For this reason, it is not uncommon for buildings to be reconfigured to meet changing needs. In a new building, there may be a trade-off between designing to meet current needs, or designing to facilitate future change by creating the option to reconfigure. This option is explored with the example of a new office building.

An unexpected requirement

The firm of Dennis, Elkins & Faye Architects has been working for about a year with Gaspero Power Ltd, a regional power utility in the UK, exploring their space needs and preparing designs for a new 10,000m^2 head office. The new building, outside an attractive market town, will bring together staff who are currently in many small offices scattered around the region. The project team of architects and space planners has worked with company managers, carefully studied the organisation and its prospects for growth, and has arrived at a design. Everyone agrees that it is an excellent design that not only meets the functional needs of Gaspero Power but is a symbol of their commitment to sustainability and emerging technologies.

The award-winning headquarters building for Wessex Water by Bennetts Associates, completed in 2000, was the inspiration for the fictional Gaspero Power building. The Wessex Water building was designed for possible future subdivision.

A few days after unveiling the design, Albert Jacklin, Gaspero's vice-president responsible for property operations, is summoned to a meeting with the vice-president for finance. At the meeting, Mr Jacklin hears that there is intense pressure from Gaspero's main investors to reconsider the design. They insist that, notwithstanding any other features of the design, the building must be arranged so that it can be subdivided easily. Mr Jacklin is stupefied: why would the investors in a major corporation be concerned about the detailed design of their main offices, and why would they want a design which reduces the building's efficiency? However, Mr Jacklin passes the instruction to the design team at Dennis, Elkins & Faye, asking them to review the implications of a redesign to facilitate subdivision. Dennis, Elkins & Faye are initially dismayed but, like most designers, stimulated by a fresh challenge.

After a week's intensive work, Dennis, Elkins & Faye report that they have come up with a design that could be subdivided into three self-contained offices. They think that there would be only a small impact on Gaspero's operating efficiency, and estimate that the increase in construction cost due to duplication of service cores and greater building perimeter is less than 10%. They actually prefer the new design to the old one, as it creates some attractive semi-enclosed courtyards.

Meanwhile, Mr Jacklin has figured out what the new design requirement is all about. The investors want to minimise the risk attached to their investment in Gaspero Power. The terms of the investment state that if Gaspero fails, the investors gain possession of the building. But office buildings are worthless unless they are occupied by rent-paying tenants, and it may not be easy to find a new tenant for the Gaspero head office in a moderate-sized provincial town where there are few organisations needing a large, single-user building. In the event of default by Gaspero, the lenders would prefer a building that could be reconfigured into tenancies of different sizes.

The lenders are concerned about the uncertainty of finding new tenants for a large office building, and they believe that the option to reconfigure the building into smaller tenancies will reduce this uncertainty.

Mr Jacklin understands the attraction of the option, but is it good value? The question can only be answered by comparing the cost of the option with its value. Dennis, Elkins & Faye have provided an estimate for the additional cost of modifying the building to create the option, and Mr Jacklin thinks it is worth spending some effort to work out the value of the option. He is not happy that the

Commercial buildings have to adapt quickly, often radically, because of intense competitive pressure to perform, and they are subject to rapid advances that occur in any industry. Most businesses either grow or fail. If they grow, they move; if they fail, they're gone. Turnover is constant. Commercial buildings are forever metamorphic.

Stewart Brand *How Buildings Learn* (1994)

investors' desire for flexibility is being taken as an absolute requirement, without being tested.

Simulating tenancies

Mr Jacklin's job with Gaspero involves leasing a variety of buildings around the region for service and maintenance staff, so he has frequent contact with commercial estate agents. He picks their brains to build up a picture of the prospects for future lettings of the Gaspero building.

It is evident that there are many more small businesses (as measured by the amount of floorspace that they rent) than large ones. The agents are a bit vague on the precise numbers. Their general observation is confirmed at the national scale by the statistics of business rates, as commercial property tax is termed in the UK: the size breakdown of rated premises is published and shows a strong skew towards smaller premises. The lack of precise data, which is familiar in the real estate world, means that Mr Jacklin cannot embark on precise mathematical analysis of the Gaspero rental problem. Instead he decides to use simulation.

This is a typical real estate problem, with inadequate data to define the 'correct' probability distribution, so Mr Jacklin decides to simulate the size of prospective tenants with the relatively simple lognormal distribution: nothing would be gained by taking a more complex approach. His lognormal distribution has an average value for tenancy size of 4,000m^2 and generates a series of simulated tenancies in which the size range 2-4,000m^2 is much more common than 8-10,000m^2 or more.

A second point made by the commercial agents is that larger firms tend to stay longer as tenants than small ones: some small firms are durable, but many are newly established and quickly grow and relocate – or disappear. Mr Jacklin simulates this by attaching a 'life expectancy' to each tenancy which is generated in this simulation. In the absence of quantified data he does this in a rough-and-ready way, each tenancy's lifetime being randomly generated around an average which is proportional to the

The Wessex Water building provides a high quality working environment with a high level of energy efficiency.

Simulation of tenancies

technique 9.1

The simulation generates prospective tenants for the Gaspero building, defined in terms of the amount of floor area desired, and the duration of the tenancy.

It is assumed the distribution of tenancy sizes follows a lognormal distribution, so that the number of prospective tenants wanting 2-4,000m^2 is far greater than the number wanting 10,000m^2. The mean size is about 4,000m^2. The simulation randomly generates floor area desired by a prospective tenant in accordance with the lognormal distribution.

Next, it is assumed that the tenancies of larger businesses are likely to be longer-lived. The average duration of a 10,000m^2 tenancy is assumed to be 20 years and of a 4,000m^2 tenancy five years – average duration being directly proportionate to size. The duration of a particular tenancy is randomly generated around the average corresponding to its size, with a minimum of two years.

The following is a typical series of twelve tenancies generated in this way. There are more small than large tenancies, and the larger tenancies tend to have a longer duration, but there is a considerable amount of random variation.

Tenancy size (m^2)	*Tenancy duration (years)*
2858	7.5
3112	8.75
5607	18.75
2627	9.25
5263	20
1069	2
3363	7.25
5839	2.25
4826	13.5
11970	45.75
7085	14.25
4533	16

When it is first built all 10,000m^2 of floorspace in the new building is occupied by Gaspero. In the simulation, the duration of their tenancy is modelled with an average life of 20 years, just like any other tenant. When they vacate, prospective tenancies have to be matched against the available floorspace. It is assumed that a prospective tenant takes up a lease if its area demand is within a 10% 'loose-fit margin' of the area available. If a lease is taken up, the floorspace remains occupied until the duration of the tenancy expires, when the floorspace again becomes available.

The simulation covers 25 years in 3 month time periods, so 100 time periods are simulated. In each time period one new prospective tenant appears. A complete *array of tenancies* starts with the 10,000m^2 Gaspero tenancy in time period 1, followed by 99 simulated tenancies, each with its floorspace demand and duration.

size of the tenancy: for a 10,000m^2 tenancy the average is 20 years; for a 4,000m^2 it is 8 years. Mr Jacklin checks his simulated tenancies with the commercial agents, who think that they look about right – not a ringing endorsement, but the best he could expect.

In the next stage of the simulation the tenancies are assigned to the Gaspero building, with its original single occupant design and with the revised design that has the option to reconfigure.The redesigned layout has three potential subdivisions, A (4,500m^2), B (3,000m^2) and C (2,500m^2), offering many alternative office sizes.

With a 10% 'loose-fit margin', the non-option building can accommodate tenants whose demand varies from 11,000m^2 to 9,000m^2, but we know that there are not many tenants of this size: whereas with the option it can accommodate most tenants whose demand varies from 11,000m^2 to 2,250m^2. However, the range of alternatives available for new tenants is reduced if part of the building is already taken by other tenants.

Because the subdivisible building can accommodate a wider range of prospective tenants, and because there are more small tenants than large tenants, the simulation shows that the offices in the subdivisible building are much more easily let. Even though small tenancies have shorter life expectancies, the smaller offices are quickly re-let. When the building is in single-tenant form there can be quite a long wait for a large tenant, with periods when the building is empty.

In the simulation, it is assumed that there is no cost attached to exercising the option to reconfigure – tenants can take one, two or three adjacent parts of the building without incurring any expenditure for adaptation.

Mr Jacklin runs the simulation a number of times and on average he finds that the building achieves 92% occupancy when it can be let as three offices, but only 84% occupancy when it has to be let to a single tenant. These are high occupancy values for a speculative office, but the Gaspero building starts off with full occupation by a large tenant – a developer's dream.

Tenancy sizes	
WITHOUT OPTION	10,000m^2
WITH OPTION	
A + B + C	10,000m^2
A + B	7,500m^2
B + C	6,000m^2
A	4,500m^2
B	3,000m^2
C	2,500m^2

Possible tenancy size without and with the option to reconfigure.

The value of commercial property

principle 9.1

The value of commercial property derives primarily from the expected cashflow of rental income from commercial tenants. All other costs faced by the owner of a commercial building are negative – outlays for construction, maintenance, fuel, cleaning, security, legal fees, taxes, etc. The owners of commercial buildings do not actually occupy them, but regard them as a form of investment – an alternative to government securities, shares or alternative investments. The value of a commercial building is therefore set by the market on the basis of expected income and risk. An empty commercial building is a disaster, unless the owner expects that a rent-paying tenant will turn up within a reasonable time.

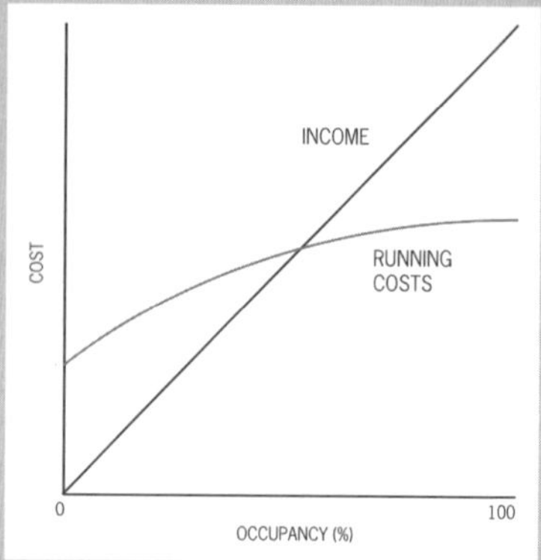

The ideal scenario for a building owner is to have a tenant or tenants fully occupying the building, paying a good rent, on a long lease with a low probability of defaulting. A strong cashflow with minimal risk creates a high market value for the building.

The reality is usually different, with tenants coming and going, and looking for the best deals from competing building owners. A priority for owners is to achieve high occupancy rates, bringing high rental income: a building that is 80% occupied will generate twice as much rent as a building that is 40% occupied.

In practice the profitability to the owner is not precisely proportional to the occupancy level, as some of the owner's costs are fixed regardless of occupancy, whilst others are only incurred when tenants are installed. The basic maintenance and security costs are much the same for a half- or fully-occupied building; the same may apply to heating and cooling costs if the plant is centralised. In general, a building that is nearly full costs virtually as much to run as a building that is completely full, so owners are delighted to gain the extra rental income from the last few percentage points of occupancy – moving from 80% to 90% occupancy, or from 90% to 100%.

It might be argued that the value of a commercial building is based not just on rental income but also on the value of the land it occupies, which could be sold for redevelopment. However, the investor in redevelopment is solely motivated by the future cashflow of rents from the replacement building, so commercial land value is itself founded on expected rental income.

Simulation of ocupancy

technique 9.2

When an array of tenancy sizes and durations has been simulated, it is assigned to the Gaspero building. There is one prospective tenant for each of the 100 time periods in the simulation, and in each time period the prospective tenant is compared to the available space, both without and with the option to reconfigure into three smaller parts.

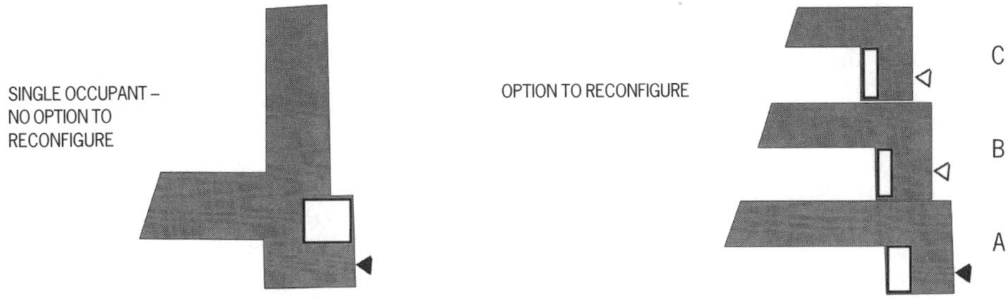

When the building is fully occupied, the new tenant cannot be accommodated.

If the building is empty and there is no option to reconfigure, it is necessary to wait until a tenant turns up who requires the whole building – this can involve waiting for several time periods. With the option, if some or all of the building is empty and the new tenant's target size is within 10% of the floorspace in a vacant part of the building, that part is assigned to the tenant, who remains in occupation for the duration of the tenancy. Because there are more small tenancies, the periods of waiting for suitable tenants to arrive are shorter with the option to reconfigure.

Typical profiles of occupancy without and with the option to reconfigure are shown in the following diagram, for one typical array of tenancies. Time runs from left to right; the shaded cells indicate that the building is occupied and the blank cells indicate that the building is empty.

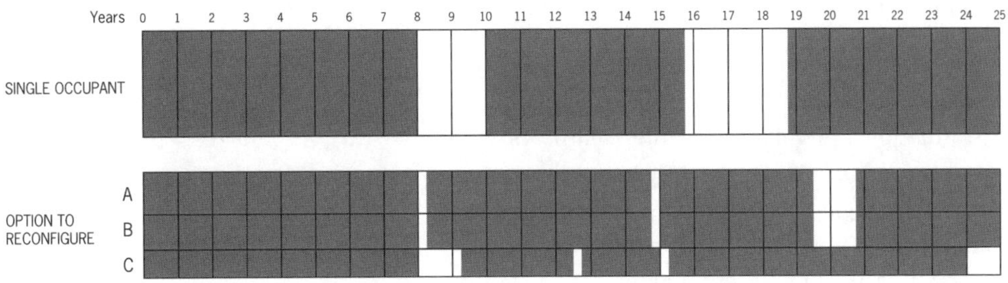

Note that without the option to reconfigure there are longer periods of waiting, when the whole building is empty. In this example, the overall occupancy is 76% without the option to reconfigure and 92% with the option.

Taking 25 simulated arrays of this type, the average level of occupancy was found to be 84% without the option and 92% with the option – an increase in occupancy of 8%. In a few cases the non-option building achieved higher occupancy, as large tenants may be turned away from the multi-tenancy building if it is already partially occupied by smaller tenants.

Simulated option value

When comparing the Gaspero building in its original single-occupant design and in the revised form with the option to reconfigure, the increase in cost for providing the option is about 5%, and the expected increase in occupancy rate is about 8%. Since an increase in occupancy translates into an increased rental cashflow, it seems that option value slightly exceeds the option cost.

A more thorough analysis of the simulation findings involves calculating the net present values of the rental cashflows, which takes account of the timing of the periods when the building is occupied or vacant during the 25-year simulation period. The cashflow for each simulated assignment of tenants is generated by assuming that tenants pay rent at £185/m^2 per year when the building or part of the building is occupied (this is a real cashflow in which inflationary rent increases are ignored).

Based on the commercial property market, Mr Jacklin believes that a discount rate of 8% per year is about right. When the simulated cashflows are discounted the average NPV for the building in single occupant form is £15.77m, and with the option to reconfigure it is is £16.78m, giving an option value of £1.01m: this is an increase of about 6.5%. The option to reconfigure still looks like good value.

However, an important difference between the simulated cashflows has not yet been taken into account. In the Gaspero simulation the volatility of the cashflows with the option to reconfigure is significantly lower than when the building is in single-occupant form – volatility in the 25 simulations is 7.5 % with the option and 18% without the option. As described in earlier chapters, the volatility of the return on an investment is what determines the risk premium. Thus the risk premium should be lower with the option to reconfigure.

The market-based 8% per year discount rate represents a 5% risk premium over the 3% risk-free rate. If the risk premium is reduced to 3% (and a greater reduction could be considered), the discount rate for the option-based rental cashflows is reduced to 6% per year. The impact

Value of the option to reconfigure

technique 9.3

It is assumed that when all or part of the Gaspero building is occupied in a simulated assignment, the tenants pay rent at £185/m^2 per year; this gives the rental cashflow for that assignment (in real tems). The cashflow is discounted to establish its net present value. The difference between NPVs without and with the option to reconfigure gives the value of the option. The results of 25 assignments are shown in the table; many more assignments, perhaps 1000, would be carried out in a Monte Carlo simulation.

	SINGLE OCCUPANT		OPTION TO RECONFIGURE		
Simulation	*Floorspace occupancy rate*	*NPV of rent at 8% discount (£m)*	*Floorspace occupancy rate*	*NPV of rent at 8% discount rate (£m)*	*NPV of rent at 6% discount (£m)*
1	88%	16.06	93%	16.97	20.21
2	100%	17.78	100%	17.78	21.30
3	73%	14.79	80%	15.83	18.53
4	88%	16.98	94%	17.35	20.63
5	97%	17.60	97%	17.63	21.05
6	38%	10.91	86%	16.36	19.32
7	98%	17.08	98%	17.08	20.57
8	96%	16.44	74%	13.40	15.98
9	95%	17.48	96%	17.54	20.91
10	39%	9.32	93%	16.51	19.77
11	65%	14.93	98%	17.59	21.03
12	100%	17.78	100%	17.78	21.30
13	67%	13.02	83%	15.85	18.70
14	88%	15.84	92%	16.80	19.98
15	100%	17.78	100%	17.78	21.30
16	82%	13.70	92%	16.35	19.56
17	93%	16.49	86%	15.90	18.84
18	100%	17.78	100%	17.78	21.30
19	69%	14.53	90%	16.73	19.85
20	100%	17.78	100%	17.78	21.30
21	87%	16.29	91%	16.76	19.90
22	91%	17.03	93%	17.18	20.40
23	99%	17.57	99%	17.57	21.05
24	79%	14.84	94%	17.00	20.28
25	68%	14.49	78%	14.32	16.99
Average	84%	15.77	92%	16.78	20.00
Volatility	18%		7.5%		

The table gives NPVs using a discount rate of 8% per year (in real terms) for commercial offices. For the 25 typical simulated assignments, the average NPV without the option is £15.77m and with the option to reconfigure it is £16.78m. The option value is £16.78m – £15.77m = £1.01m.

However, note that the occupancy levels in the simulations are much less volatile with the option to reconfigure: this justifies a lower risk premium. If the discount rate is dropped from 8% to 6% per year for the rental cashflows with the option, the average NPV rises to £20.00m. Cashflows without the option are still discounted at the higher 8% rate, so NPV without the option remains £15.77m. The option value is now £20.00m – £15.77m = £4.23m.

on NPV is dramatic. With 6% discount rate, the average NPV of the option-based rental cashflows rises to £20.00m in the 25 simulations. When this is compared with the average NPV of the non-option cashflows discounted at 8%, which is £15.77m, the value of the option to reconfigure is £4.23m. The option creates an increase in NPV of 27%.

Commercial agents prefer to let buildings with a wide variety of tenancy sizes, to attract as many tenants as possible.

It appears that Gaspero's investors were right to insist on the option to reconfigure, which can be created by a redesign which adds about 5% to the building cost. Preliminary simulations suggest that the average occupancy increases by about 8%, but the value of the option is greater, adding about 27% to the NPV of expected rental incomes. The option may appear to have little value to Gaspero, as its primary purpose is to protect the investors if Gaspero fails, but there are also many possible scenarios when the flexibility of the option to reconfigure could be exploited by Gaspero – if, for example, the company wanted to sub-let some space or relocate its head office. In the shorter term, the option may help Gaspero negotiate a lower interest rate on its loan. Mr Jacklin is glad that the need for the option was identified, even at a rather late stage in project development.

The preliminary simulation should be followed up with careful testing of the assumptions and more simulation runs, but it is important to note that the option to reconfigure is valuable for two reasons: first, it increases occupancy and hence rental income, and second, it reduces volatility and hence the risk premium. These two factors greatly increase the expected NPV. In this example, the impact of reduced volatility is much more significant than increased occupancy.

Alternative approaches to risk management

There are alternative ways for the investors to protect themselves against the risk that their investment in Gaspero might perform badly. They could purchase some sort of credit derivative which would pay out should Gaspero fail, or some other event occur leading to reduced credit-worthiness of the company. Credit derivatives allow investors to manage risk when they make a loan: the premium they pay for someone else to assume the risk will relate to the perceived riskiness of the loan.

One might compare the cost and benefits of acquiring a lifecycle option or buying a credit derivative. Unfortunately, the pricing of credit derivatives is not well developed: various institutions use different models to price credit risk and the markets are illiquid, so it is hard to observe how the market prices this type of risk. In a real estate or building context, the risk extends over a considerable length of time – considerably longer than the life of most credit derivatives.

It may be an avenue worth pursuing, but not by Mr Jacklin. However, the possibility of Gaspero defaulting cannot be dismissed. Major corporate collapses sometimes occur with little advance notice. Moreover, whether or not Gaspero Power fails may be entangled in corporate politics, finance, and any curious activities in which the management happens to indulge. Is the failure of Gaspero Power likely? Initially one might think not, but remember that there used to be a major energy company called Enron.

There is an analogous choice between ways of managing risk in the context of design-and-manage projects. Consortia, usually from the private sector, undertake to construct a building for a client who is usually in the public sector, and manage it for a fixed period, often 25 or 30 years. The consortia take on risks associated with the uncertainty about building use over 25 or 30 years, and as a risk management strategy they often draw up the tightest possible limits on the functionality which must be supported by the building. Then, if the user needs stray outside the defined limits, the consortia can seek additional payments to adapt the building. Drawing up the contracts to limit functionality is time-consuming and expensive in legal fees. An alternative risk management strategy would be to create lifecycle options in the building, increasing the range of functionality. This would reduce risk to the consortia and increase usefulness to the client.

Instances of reconfiguration

The benefit of the option to reconfigure is well understood in speculative office design, because uncertainty about future tenants is central to speculative development. The idea is not expressed in the options terminology proposed in this book, and designs are evaluated on the basis of gut feel, or expert judgement, rather than being quantified. Extending the principle of reconfiguration to owner-occupied buildings is less well-established, because the clients' and designers' attention is focused on the needs of the commissioning organisation. The option to reconfigure was a design requirement for the Wessex Water offices designed by Bennetts Associates near Bath, where the option has not yet been exercised. A counter-example is the owner-occupied Ionica building designed by RH Partnership in Cambridge (see p.36), which was not designed with an option to reconfigure, and was difficult to re-use when Ionica went bankrupt five years after they moved into their new headquarters.

Within office buildings the use of demountable partitioning creates the option to reconfigure floor layout quickly and cheaply; the option is created by paying more for demountable partitions than for studwork and plasterboard partitions, which are cheap to build but slow, messy and expensive to demolish and rebuild. The fact that some demountable partitions are never moved does not prove that there was over-investment – but there may have been. Unless option value is estimated, it is very difficult to make a rational decision on whether to buy a lifecycle option.

The option to reconfigure helps to manage uncertainty and is particularly appropriate where change is expected. When a department store chain wanted to redevelop their city-centre site in Cambridge they built a new temporary department store in a non-central location: the temporary store is designed as a permanent building which has the option to be reconfigured as individual ground floor shop units with offices and apartments on the upper floors. Here the option maximises the re-usability of the building after the planned move of the department store back to

A new commercial building for temporary occupation by a department store, which was built with the option to reconfigure. When the department store moves out, the building can be reconfigured to provide ground floor shops, upper floor offices, and penthouse flats.

its city-centre site, but who knows whether events will unfold as presently expected, or whether the option to reconfigure will be exercised in unforeseen ways?

The option to reconfigure is encountered in residential buildings. An example of the option being identified in a situation where it might seem to have a low priority is the first generation of public housing in Hong Kong, built in the 1950s. Because of a severe problem of squatter settlements, the government of Hong Kong started an emergency housing programme to re-house as many families as possible as quickly as possible, in basic but sanitary conditions. The new housing was in six storey blocks, in which each family had a room of $11m^2$ opening off a shared perimeter balcony, giving minimal privacy. Even in the emergency situation the designers realised that the housing blocks had an option to reconfigure, by joining back-to-back $11m^2$ units into larger units, accessed from the shared perimeter balcony on one side, with the balcony on the other side becoming private. This option to reconfigure was exercised in many blocks, but because of rapid improvements in the economy and housing standards in Hong Kong, others were demolished and replaced by better (and higher density) housing before the option was exercised.

A more elaborate approach to the option to reconfigure housing has sometimes been proposed. An example from the 1970s was the PSSHAK scheme – 'primary support structures and housing assembly kits'. The idea was to build houses with permanent structural elements and changeable secondary elements, thus creating many options to reconfigure. A development of 44 PSSHAK dwellings was built in London in 1976, and a recent survey investigated the take-up of the option to reconfigure over a period of nearly 30 years – although this is a relatively short period in the life of UK houses. The findings of the survey seemed disappointing. Fewer than a quarter of the respondents had made alterations, and of this group fewer than a third believed that the PSSHAK strategy had made the alterations easier. A majority of respondents were not aware that the dwellings had been specially designed to facilitate reconfiguration. Lack of information,

The kit-of-parts idea lives on, as in a recent fire station in Hiroshima, Japan (Riken Yamamato, architect). The neutral steel grid is meant to create the lifecycle option to add or remove secondary elements. The image of reconfigurability may be more important than its practical use.

and the buildings' rather conservative appearance, may have inhibited the residents' take-up of the option to reconfigure; but it also leads one to question the value of the option that was created in the PSSHAK approach. Perhaps there was over-investment in the option to reconfigure, but this could only be revealed by a thorough appraisal of the costs and benefits and other information available to the decision-makers in the 1970s.

Kit-of-parts approaches to the option to reconfigure are still encountered. A recent proposal by John Fernandez of MIT involved buildings made of components with 'diversified lifetimes'. Over time, components would be demounted and moved using an 'array of integral cranes and trolleys' to an on-site store, where they would be available for re-use elsewhere in the building. The strategy certainly creates an option to reconfigure, indeed it would dominate the design. However, as for all lifecycle options, before investing in the option it would be prudent to assess both the cost of creating it and its option value, to avoid under- or over-investment.

Summary of the option to reconfigure

The reconfiguration of old buildings is extremely common: there are probably very few old buildings that have not been reconfigured to some degree. In most cases, the original designers made no special provision for reconfiguration, but it happened anyway. The same is likely to apply to new buildings: even if future reconfiguration is ignored by the designers, it will happen – perhaps with expense and disruption that could have been avoided.

Providing the option to reconfigure is an attractive response to this situation. Designers have put forward a huge variety of ideas for creating the option to reconfigure. Some involve high-tech solutions with demountable components whereas others provide the option through a basic building layout which allows many alternative configurations with a nil or small exercise cost.

Designs which positively inhibit the option to reconfigure risk early obsolescence: for example, a design with

The Queen's House in Greenwich by Inigo Jones, the first fully classical building in Britain, was begun in 1617, but even before it was finished the layout started to change. Many alterations of the eighteenth and nineteenth centuries were in turn stripped out in the 1930s – this can be seen as another episode in the building's history of reconfiguration.

loadbearing partitions enclosing workgroups in a commercial office. There are anecdotal examples of Victorian buildings that were built so massively, perhaps with the intention of ensuring a long building life, that they are almost impossible to adapt for changing needs.

Whenever the option to reconfigure is identified as a possible strategy it is essential to estimate the cost of acquiring the option and the value that it brings.

10

THE OPTION TO REFURBISH

Many buildings operate in ways that were not anticipated by their designers, due to changes in the surrounding environment or the activities of the users. Buildings are refurbished in response to these changes, but because the world is continuing to change, people who are considering refurbishment face uncertainty about the long-term value of their investment. Lifecycle options are relevant to the management of this uncertainty. Refurbishment at the Chauncey Club is taken as a case study.

To refurbish or not to refurbish

In earlier chapters we examined the deliberations about whether to invest in the refurbishment of a residential block at the Chauncey Club, the conference centre owned by Kingston Oil. The company, advised by their experienced property manager Horatio Black, decided to proceed. The project team are now considering specific details of the work to be undertaken, and are encountering further difficulties where there is a mismatch between the manager's gut feel and what the numbers seem to say.

Horatio Black, who has many years of experience in both constructing and managing buildings, has serious concerns about proceeding with some individual elements of the refurbishment – in particular the energy-saving aspects of the project. The design team has included the replacement of the ageing central heating boiler and single-glazed windows, even though they are still functional. Mr Black is not convinced. He agrees that wherever the walls and roofs have to be opened up, new insulation should be inserted, but he is sceptical about making energy-saving changes to parts of the building that could otherwise be left alone. This is a problem: the design team does not understand Mr Black's caution; and as an energy company

The Chauncey Club refurbishment involves upgrading a Victorian villa like this one, which offers many refurbishment options.

Kingston Oil has a policy of being a good steward of resources. To refurbish one of their own facilities and ignore energy-saving opportunities seems to conflict with corporate philosophy.

When pressed, Mr Black explains that in his experience this type of investment, focusing specifically on energy saving, does not seem to have a good track record in achieving promised benefits. His own rule of thumb is to use a payback hurdle of three years for energy-saving investment, unless there are other benefits, such as increased comfort which would increase the marketability of the Chauncey Club.

First steps in analysis

Horatio Black has expressed his performance criterion for energy-saving investment in terms of the payback period. In Chapter 2 we suggested that this is a crude method of analysis.

A better method of analysis is standard whole-life costing, which establishes the net present value of an investment. Nigel Frost, the project engineer from Blackthorpe & Partners, services consultants for the Chauncey Club refurbishment, decides to carry out a standard WLC analysis. Nigel is young and enthusiastic, and this is the first project for which he is the lead engineer, so he wants to test Mr Black's gut feel with some rigorous analysis.

Nigel Frost assembles the cashflow data for the standard WLC calculation. The initial outlay is the estimated capital cost of the replacement of the boiler and windows, which is £34,000. The annual benefit results from lower gas consumption due to a more efficient boiler and lower heat losses through double-glazed windows; the monetary saving is calculated with current gas prices, and comes to £3,240 per year. The discount rate is 8% per year (in real terms), which was originally set by Mr Black for the refurbishment project. Nigel uses a 25-year WLC study period, beginning in Year 0 with 24 years of savings from Year 1. The calculation shows a present value of savings of £34,113, and therefore positive NPV of £113, so replacement appears to be a good investment – contrary to Mr Black's gut feel.

A meeting is arranged. Mr Black agrees that Nigel Frost has used correct data for today's situation, but in discussion it emerges that he has four reasons for querying Nigel's standard WLC study as a basis for assessing long-term value.

First, gas prices, which Nigel has assumed will remain constant in real terms. However, the price of gas is volatile and in the UK (as in North America) there was a falling trend over the 20 years from the early 1980s, when Mr Black was gaining experience in the construction industry. The falling price meant that most energy-saving expenditures made during that period, if based on a projection of stable energy costs, produced disappointing results. Mr Black does not have confidence in any assumption about constant gas prices, and is unwilling to make any sort of gas price projection.

Second, new technology, as proposed for the high-efficiency boilers, does not always deliver what is promised. Maybe the technical press makes too much fuss when high-profile energy-saving projects are shown to under-perform, but Mr Black's own experience of problems with new technology makes him cautious. Building managers sometimes find that the cost of maintaining new technologies can be higher than the savings.

Third, effectiveness, in that the boilers and windows may be old and inefficient, but they do work – Mr Black thinks they could keep going for another 10 years. Hotel guests have no interest in how old the boiler or windows might be so long as the building is warm and the water hot, whereas they are very sensitive to en-suite rather than shared bathrooms. Old boilers and windows have little or no impact on the fundamental conference centre business.

Finally, risk, which is in some ways a restatement of the other points. The level of risk attached to the flow of benefits from the energy-saving investment is much greater than for the overall refurbishment project. The market niche in which the Chauncey Club operates is mature and reasonably stable, and the discount rate that was used

Cambridge University has chosen to hold rather than exercise the option to replace the single-glazed steel windows at the Fenners sports centre (see Chapter 3), despite their energy-inefficiency.

for project evaluation reflects this. It is not an appropriate discount rate for an energy-saving investment – a higher figure should be used. One discount rate does not fit all decisions, even within one corporate setting or a single project.

Mr Black believes that the new boiler and windows are much less attractive than indicated by Nigel's standard WLC calculation.

Standard WLC with risk-adjusted discount rate

Nigel Frost's first response to his realisation that the investment in energy saving is riskier than the overall investment in the Chauncey Club refurbishment, is to alter the discount rate in his standard WLC analysis.

WLC requires a risk-adjusted discount rate. In Chapter 4 the discount rate for the Chauncey Club refurbishment was established at 8% per year: it is made up of the 3% risk-free rate plus a 5% risk premium which reflects the level of risk in the hotel industry. Because investment in energy saving is riskier, a higher risk premium is needed. The premium is proportional to the amount of risk, which is typically measured by the volatility of the returns from the investment. When the volatility of a 20-year dataset of gas prices (and therefore the value of savings due to reduced gas consumption) is analysed, it turns out to be just about double that of the hotel industry. Therefore the risk premium should be doubled, from 5% to 10%. This means that the risk-adjusted discount rate for energy savings should be 13% per year, that is, the 3% risk-free rate plus a 10% risk premium. These are all real rates, which ignore price rises due to inflation.

Nigel re-runs the standard WLC analysis with the higher risk-adjusted discount rate of 13%, for 24 years of savings at £3,240 per year. The present value comes to £23,596, much less than the present value of £34,113 which he obtained with the 8% discount rate – confirming that the discount rate is critical for WLC analysis. The capital cost estimate for the new boiler and windows is £34,000, so the net present value is negative and the investment is not attractive.

Using historic data to calibrate the binomial tree

technique 10.1

Good quality historic datasets are hard to come by in the construction and real estate sectors, in complete contrast to financial markets where they are plentiful. In the few cases where datasets are available, they should be used to calibrate the forecast of future uncertainty. Historic fuel prices are one of the few sources of good time-series data..

The table gives the gas price index in the UK, in real terms (inflationary price rises stripped out), for a 20-year period:

	Index of gas prices in real terms (1990=100)	*Annual % variation*
1984	171.8	
1985	173.6	1.05%
1986	146.7	-15.50%
1987	131.1	-10.63%
1988	117.6	-10.30%
1989	105.7	-10.12%
1990	100	-5.39%
1991	94.7	-5.30%
1992	94.4	-0.32%
1993	90.3	-4.34%
1994	89.8	-0.55%
1995	76.3	-15.03%
1996	53.9	-29.36%
1997	55.3	2.60%
1998	56.7	2.53%
1999	55	-3.00%
2000	59.2	7.64%
2001	77.8	31.42%
2002	71.8	-7.71%
2003	72.5	0.97%
2004	79	8.97%
Average	94.0	-3.12%
Variance	1314.5	1.44%
Volatility	36.3	12.01%

To construct a binomial tree, the ratio which determines the rising and falling values at each branch has to be set. The appropriate figure is the volatility of the annual variation, which in this case is 12%. Note that in a few years on the period 1984-2004 the annual variation exceeded this figure; but a higher figure for annual variation would cause an excessively wide spread for the binomial tree. At each branch in the binomial tree:

$$e_u = e_i \times (1 + w) \qquad e_d = \frac{e_i}{(1 + w)}$$

where e_i = the current energy price
e_u, e_d = the prices on the rising and falling branches
w = the rate of change.

Nigel wonders whether the expected life of 24 years may be an underestimate, so he re-runs the analysis with an expected life of 34 years. The present value only increases to £24,532. This demonstrates that high discount rates lead to a short-term perspective, in which costs and benefits occurring far in the future carry very little weight.

Nigel shows his new findings to Mr Black, who is gratified that his gut feel has been confirmed, but he doesn't dismiss the idea of replacing the boiler and windows out of hand. He makes the observation that he will probably make the replacements if the price of gas goes up in the future. These are simple retro-fit upgrades that can be done during the conference centre's low-booking period from July to September. Nigel Frost realises that Mr Black has the option to replace the boiler and windows in the future, until they collapse when replacement cannot be put off any longer. This is in contrast to the ensuite bathrooms that must be installed now, in order to ensure the survival of the conference centre business. Nigel sees that this lifecycle option could make a big difference to Mr Black's willingness to invest.

Binomial tree of gas price uncertainty

Nigel Frost proceeds to check out some textbooks on real options and gradually comes to grips with Mr Black's decision environment, which includes the lifecycle option to refurbish the boiler and windows. The lifecycle option is attractive because of uncertainty about the future gas price, and the value of savings that would result from reducing gas consumption. With the option, Mr Black can wait and see whether the price of gas rises or falls; if it rises sufficiently he will exercise the option and replace the boiler and windows, otherwise he will prefer to hold the option until the components fail and have to be renewed. Meanwhile, Mr Black has the benefit of holding a lifecycle option: he can make an effective investment if the opportunity arises, but faces little risk of making an investment that delivers poor returns.

Present value of a binomial tree of cashflows

technique 10.2

It is mathematically straightforward to establish the present value of a cashflow of estimated values: each entry is discounted with the risk-adjusted discount rate for each time period between now and the time when the entry occurs.

$$\text{present value} = v_0 + \frac{v_1}{(1 + r)} + \frac{v_2}{(1 + r)^2} + \ldots \frac{v_n}{(1 + r)^n}$$

where v_0, v_1, v_2, ... are the estimated cashflow entries in time periods 0, 1, 2, ... and r is the risk-adjusted discount rate. In the energy-saving refurbishment example there is a high risk-adjusted discount rate as the estimated cashflow of future savings is uncertain.

When the estimated cashflow of future savings is replaced with a binomial tree of possible savings the calculation is more complex. It can be illustrated with a single binomial branch: discounting a binomial tree is just a matter of discounting every branch in the same way.

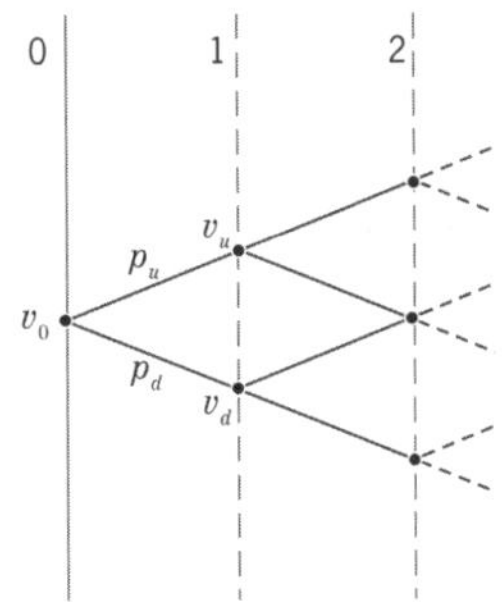

The cashflows from rising and falling branches are discounted with the risk-adjusted discount rate and weighted for the probabilities of rising and falling:

$$\text{present value} = v_0 + p_u \times \frac{v_u}{(1 + r)} + p_d \times \frac{v_d}{(1 + r)}$$

where p_u is the probability that the rising cashflow v_u will occur and p_d is the probability that the falling cashflow v_d will occur; in all cases $p_u + p_d = 1$.

This calculation is done at each branch in the binomial tree, starting on the right-hand side. The present value is the result of the last calculation.

If the binomial tree is symmetrical, with no long-term trend of rising or falling values, then (i) the rising and falling cashflow values are generated from the same change ratio w:

$$v_u = v_0 \times (1 + w) \quad \text{and} \quad v_d = \frac{v_0}{(1 + w)}$$

and (ii) the rising and falling probabilities are equal:

$$p_u = p_d = 0.5$$

and the present value of the binomial tree is approximately the same as the present value of a cashflow with all entries equal to v_0. The lower the change ratio w the closer the approximation. The primary reason for constructing a binomial tree is not to establish the present value of an uncertain cashflow, but to explore the value of lifecycle options.

Valuing the option to refurbish technique 10.3

The option value calculation starts with the binomial tree of annual savings that would result from replacing the boiler and windows at the Chauncey Club. This is the beginning of the tree that runs from Year 0 to Year 24:

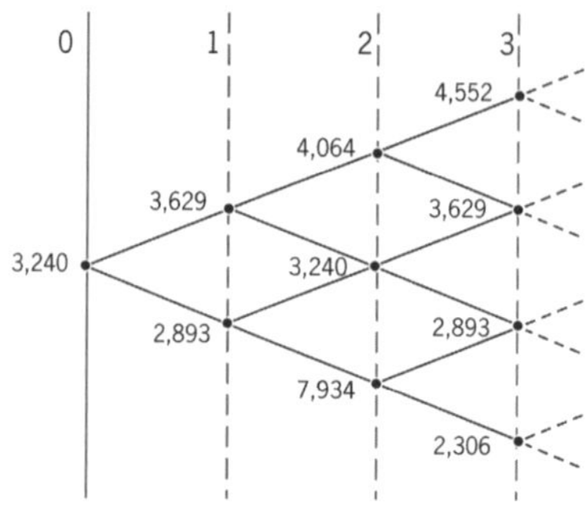

Consider the branch in Year 3 where the savings are £4,552 per year. The corresponding branches for the present value of 24 years of savings, and NPV if the exercise cost is £34,000 are:

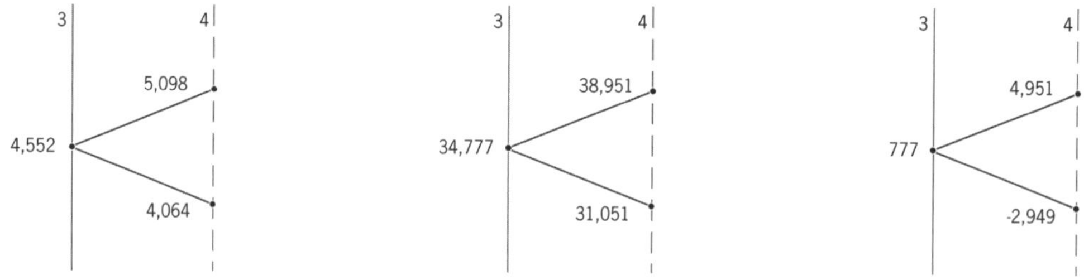

Savings Present value Net present value

If the option to refurbish is exercised in Year 3, it has an NPV of £777 and two possible outcomes in Year 4:

Rising gas price	*Falling gas price*
Saving of £5,098 between Years 3 and 4	Saving of £4,064 between Years 3 and 4
PLUS present value of 23 further years of savings, £38,621	PLUS present value of 23 further years of savings, £30,789
£5,098 + £38,621 = £43,719	£4,064 + £30,789 = £34,853

Alternatively if the option is held in Year 3 there are two possible outcomes in Year 4:

Rising gas price	*Falling gas price*
project with NPV of £4,951	project with a negative NPV of –£2,949, and therefore zero value, £0

How can the value of holding this option be calculated? An ingenious method is via a *matching portfolio* which has exactly the same outcomes as building the option. The matching portfolio may not be practicable in reality, but its hypothetical performance and price can still be established.

The matching portfolio involves investing in a smaller proportion of the risky project, partly financed by borrowing cash. Suppose the proportion is 55.8% and £18,533 is borrowed for one year at a 5% interest rate. The matching portfolio has two possible outcomes in Year 4:

Rising gas price	*Falling gas price*
asset with present value of £24,410 (55.8% x £43,719)	asset with present value of £19,459 (55.8% x £34,852)
LESS obligation to repay the loan which has risen in value to £19,459 (1.05 x £18,533)	LESS obligation to repay the loan, £19,459
£24,410 – £19,459 = £4,951	£19,459 – £19,459 = £0

These are the same outcomes as for holding the option. The cost of acquiring the portfolio is 55.8% of the present value of the project in Year 3, 55.83% x £34,777 = £19,417, less the loan, £18,533, leaving an outlay of £884. Therefore, this must be the value of holding the option in Year 3. The value of exercising the option is £777, so it is better to hold the option.

The formulae for calculating option value in this way are:

$$\text{matching ratio } H = \frac{(T_u - T_d)}{(B_u - B_d)} \qquad \text{option value } Q = H \times V_i - \frac{(H \times B_u - T_u)}{(1 + r)}$$

where

T_u and T_d = value of exercising the option in Year i +1 on the rising and falling branches, or 0 if the value is negative; here, £4,950 and £0.

B_u and B_d = benefit on the rising and falling branches in Year i+1 if the option is exercised in Year i: that is the benefit in Year i+1 plus the present value of future benefits; here, £43,719 and £34,853 as already established.

V_i = present value in Year i; here, £34,777

r = interest rate for borrowing; here, 5%

In the example:

$$H = \frac{(4951 - 0)}{(43719 - 34852)} = \frac{4951}{8867} = 0.558, \text{ or } 55.8\%$$

$$Q = H \times 34777 - \frac{(H \times 43719 - 4{,}951)}{1.05} = 19417 - 18533 = 884, \text{ or £884.}$$

This is the value of a one-period option; the value of a multi-period option will be higher – it is established by applying the method at each branch of the binomial tree, starting on the right and working back to Year 0. Notice that the matching portfolio method does not require a risk-adjusted discount rate.

A mathematically equivalent way of calculating option value is the 'risk-neutral' method, which is described in options and real options text books. Both methods produce the same option value.

To establish option value, Nigel has first to construct a binomial tree of future gas prices. His analysis of the volatility of historic gas prices leads him to use a ratio of 12% to establish the rising and falling prices at each branch of the binomial tree. The tree begins with the current gas price in Year 0, and the fan of rising and falling gas prices continues from Year 1 to Year 24. As Nigel has no knowledge of whether gas prices will tend to rise or fall in the long term, he assumes that the rising and falling prices are equally probable.

A corresponding binomial tree shows the annual savings that would be achieved from the replacement boiler and windows, for the corresponding gas price. This starts with £3,240 at Year 0, the value of savings at today's gas price.

From this binomial tree of savings, the present value at Year 0 of 24 years of savings can be estimated, given the uncertainty captured by the binomial tree. At each branch of the tree of annual savings, the two alternative future values are weighted by the probability of occurrence (0.5 for each branch in this case) and discounted at the 13% discount rate (see p. 140). This is done at each branch and all savings are discounted back to Year 0. In this year no savings are counted as it is assumed that the refurbishment is being carried out.

This gives the expected present value of savings over 25 years if the boiler and windows are replaced in Year 0, when the annual savings are £3,240; the value is £24,750 (a little higher than the figure of £23,596 for NPV with constant savings of £3,240 per year, because the numbers on the rising side of the binomial tree grow more quickly than the falling figures decline). Suppose that the replacement is not carried out until Year 1, and the price of gas rises between Year 0 and Year 1: then the replacement would lead to savings over 25 years with initial value of £3,629 per year; this gives a present value of £27,720. The same exercise can be carried out for each node of the binomial tree.

The net present value of refurbishment is the present value of savings less the capital cost of the work, which is

Holding or exercising the option

technique 10.4

In the specimen branch described in technique box 10.3 the value of holding the option to refurbish is greater than the benefit from exercising it, so it is better to hold the option and leave the building in an unrefurbished state – this is despite the fact that exercising the option has a positive NPV, so that in the absence of an option the refurbishment should go ahead.

The decision to hold the option is specific to the prevailing data, particularly the gas price which determines the values of savings due to reduced gas consumption. In the following years the gas price will change in unpredictable ways, and the decision on whether to hold or exercise the option must be revisited.

The outcome of the decision can itself be mapped onto a binomial tree, with the typical form shown in the diagram: 'wait' means that it is better to hold the option, 'go' means that it is better to exercise it.

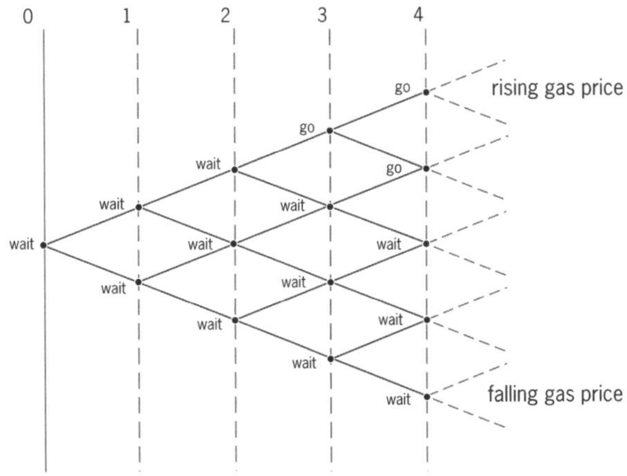

The 'go' decisions coincide with high gas prices, and therefore high savings from the refurbishment. Essentially, if you have an option, you wait until the NPV is high enough to make the risk of a bad outcome very small.

The binomial tree of possible future decisions is not a prediction of when the option will be exercised: that will depend on the flow of actual events, not on today's forecast of possible events. It is, however, important that there are some 'wait' and some 'go' entries. If there were no circumstances in which it would be preferable to exercise the option rather than hold it, there would be very little, if any, benefit in having the option. And vice versa, if there were no circumstances in which it would be preferable to hold the option rather than exercising it, the option would be worthless.

£34,000. At Year 0 the present value is £24,750 so the NPV is –£9,250, confirming the conclusion of the standard WLC using the risk-adjusted discount rate: the investment is not viable. There is an NPV for each node on the binomial tree of present values; some NPVs are positive and some are negative.

Options decision-making

Having assembled all this data, Nigel Frost is now able to quantify the impact of Mr Black's lifecycle option to refurbish.

The important point is that when the NPV is negative, Mr Black does not permanently reject the refurbishment, but waits to see what happens to the future gas price. Waiting is the same as holding the lifecycle option to refurbish. Only when the gas price rises to generate a sufficiently high NPV will Mr Black exercise the option and carry out the refurbishment.

This situation only applies for the duration of the option. When the option expires Mr Black cannot wait, but has to act. In this example, the option expires when the old boiler and windows fail.

How does Mr Black decide whether to retain the option or exercise it? When the NPV is negative he would never exercise the option. When the NPV is positive, he compares: 1) the NPV of carrying out the refurbishment now, and 2) the option value. If the option value is higher than the NPV, Mr Black retains the option – he waits.

The owners of these two semi-detached houses chose different ways of exercising their option to refurbish.

In Year 3, for example, if the gas price rises to give annual savings of £4,552 per year, the NPV of exercising the option and going ahead with the refurbishment is £777, but the value of holding the option is £884, so it is better to wait. If the gas price goes up again in the following year it will be better to exercise the option; if it falls the option will again be held.

The comparison between exercising the option and holding shows that:

- if the NPV of the refurbishment is negative, hold the option;
- if the NPV of the refurbishment is positive but small, it is still better to hold the option;
- if the NPV of the refurbishment is strongly positive, exercise the option.

Again, Nigel discovers that Mr Black's gut feel is confirmed. When there is a lifecyle option which allows an investment to be deferred, it is reasonable to require a higher rate of return from the investment before proceeding.

With his modelling assumptions, Nigel discovers that it would only be reasonable to exercise the option when the gas price had risen significantly – perhaps not the 'three year payback' that Mr Black had spoken about, but a high threshold. Perhaps other sources of uncertainty, in addition to gas prices, could push the threshold up, for example, uncertainty about boiler performance and gas usage in the refurbished building. Nigel is happy to exclude boiler and window replacement from the scope of work in the Chauncey Club refurbishment.

As always, Nigel's preliminary model should be thoroughly reviewed to check that he has made valid assumptions about source data and the method of analysis.

The energy paradox

Experienced managers are sometimes not fully aware of why they take the courses of action they do. In this case, his accumulated experience tells Horatio Black to demand a higher rate of return for energy saving investments than for other investments, which he expresses as a short 'payback period'. When the decision is more closely examined it is found that he has good reason for caution.

In the case study, the proposed energy-saving investments were set aside even though they appeared at first sight to offer a return on investment that was comparable to the refurbishment project as a whole. This phenomenon is often encountered, to the intense frustration of energy conservation advocates. As a result, governments, in order to achieve public policy objectives, are forced to undertake aggressive incentive programmes or simply require that energy-saving measures must be adopted.

Uncertainty, irreversibility and the option structure combine to make managers cautious about energy saving investments, even when they seem to offer a respectable return at current fuel prices. This 'energy paradox' has been explored by a number of researchers, who explain the rationality of the hurdle for energy-saving investments being considerably higher than might otherwise be expected, perhaps up to four times greater, because of the volatility of energy prices and option structure underlying many decisions. The uncertainty and irreversibility apply to energy saving investment in new buildings, but the option structure is particularly characteristic of refurbishment, when managers have the right, but not the obligation, to replace functional but energy inefficient components. If they are not replaced this year, they can be replaced next year, or the year after: managers can wait until patching up is no longer practical. Holding the option to refurbish is not spinelessness, it is a rational strategy for managing uncertainty and reduces the risk of over-investment. But managers should beware of continuing to hold the option long after it is 'in the money', when it should be exercised.

One might ask whether gas prices are so important. The objective of energy conservation is to save energy, not money, and the same amount of energy will be saved by the proposed energy conservation measures whether the

Major refurbishment in progress at the Canadian High Commission in London.

Organisational structure makes a difference

principle 10.1

The internal structure of the organisations participating in the research made a difference in the way they perceived energy investments. If all aspects of a conference centre are the responsibility of a manager (or team of managers), they will concentrate their efforts on what dominates the success and failure of the facility – typically staff and staffing costs, marketing, service quality and purchasing. Energy is a small part of their total budget and gets little attention. If facilities costs are the responsibility of a separate property department, energy is a big concern to them and they work on energy saving as a priority.

gas price is high or low. The reason why fuel costs are relevant is to do with opportunity costs – the benefits that could have been obtained if the financial investment had been made in other ways. Alternative energy saving investments, which from a whole-life perspective must include both capital and running costs, should be compared against the energy saved. Using a measure of, say, the energy saved per £1,000 of whole-life financial investment, alternatives can be ranked. To get the greatest energy saving from a given financial budget, the highest performing alternative must be chosen. Energy saving investments that give a high financial return will leave more resources for investment in other energy saving measures: in effect, the money savings from the first measure can be spent on the second measure.

Lifecycle options and refurbishment

Buildings that are functional but no longer in first class condition present many options for refurbishment – and this includes most existing buildings. In emergencies, when delay could be catastrophic, building owners respond quickly and get the roofer or plumber on site as fast as possible; but for non-urgent improvements there is the option to refurbish later, and most people hold this option for a long time before exercising it. Often a trigger point – such as change of ownership or use, or failure of a critical component – means that refurbishment is required, and this is when a whole bundle of options are exercised.

Even when there is a financial drive for refurbishment, market uncertainty can create problems for defining the scope of work. A study of the refurbishment of 30 year old apartment blocks in Finland found that owners were reluctant for pay for radical refurbishment, in case weak future demand for high quality apartments might depress rental levels and give a poor return on the investment. Minimal refurbishment at low cost could give the apartment block owners the option to undertake a second-stage refurbishment at a later date, if justified by market conditions. Holding the option on high quality refurbishment incurs less risk than investing under uncertain market conditions.

If the buildings of a property are rehabilitated thoroughly (at high cost) … the property owner has limited financial possibilities after rehabilitation to react to changes in the demand for the property. But, if the buildings of the property are rehabilitated to a degree that is low … ensuring nevertheless that that all damaged elements are repaired, the property owner has a wide range of possibilities [= options] with which to react to changes in the demand for the property.

A Saari *The Viability of Rehabilitated Apartment Buildings* (1994)

Many refurbishment options are embedded in existing buildings, but it is possible to create options for future upgrading in new buildings. This was proposed in relation to design for energy conservation by David Fisk in the 1970s, at the time of the first oil crisis when energy conservation first became a headline issue. Fisk suggested that the rather low regulations for thermal performance that applied to new housing in the 1970s might be superseded by higher standards during the lifetime of houses then being built. He suggested that thermal insulation measures that could only be installed at construction stage, for example, insulation below the ground floor slab, should be prioritised; but where there was an option for future retrofit, for example cavity wall insulation, the option should be created, to be exercised at some future date if it became appropriate to make the additional investment.

Many buildings have so many refurbishment options that they can be kept in service for an indefinite period by periodically exercising some of the options.

Refurbishment makes up a larger part of the UK construction industry workload than new-build, so there are innumerable lifecycle options to refurbish. When faced with an uncertain future, clear thinking about option value, and quantification when appropriate, will minimise the risk of under- or over-investment.

11

OPTIONS ON NEW TECHNOLOGY

Recent history tells that technical progress is relentless, but the rate of progress and the direction it will take is unpredictable. Investors are naturally reluctant to make expensive commitments to technologies that are likely to change drastically in a few years. Often it is more attractive to acquire the option to adopt the technology. Photovoltaic energy generation for houses is taken as an example.

An exciting innovation

The Heathland Housing Association provides much-needed social housing in a rural district where many of the old cottages have been bought by affluent Londoners as second homes. Typically the Association builds groups of six to ten houses on brownfield sites in small towns which have basic local services – shops, a primary school and public transport.

Heathland is a go-ahead organisation. It has an innovative scheme for tenant participation; it chooses sites that are too challenging for most developers; and it likes to incorporate new technical developments in its projects – previous schemes have used super-insulated timber framing and grey water recycling.

For a new site in Little Hitcham the Association's Director of Development, Jeremy Keen, would like to try photovoltaic electricity generation, or PV. (Don't confuse the PV of 'photovoltaic' with 'present value' – in this book we do not use PV for present value.)

PV gets lots of publicity and it is a seductive idea: you put high-tech cells on the roof of a house and they produce free electricity. There have been quite a few demonstration schemes, some on housing association projects, and in the long term PV is seen as a major contributor to the

To make best use of photovoltaic technology, new housing should be laid out with south-facing pitched roofs and as little overshadowing as possible, unlike typical developers' layouts.

switch from fossil fuels to renewable energy. And just now, in the UK, there are big government grants towards the capital cost of PV cells.

It sounds great, so Jeremy Keen starts putting together some budget figures for PV at the Little Hitcham site.

Is PV cost-effective?

To evaluate the PV installation at Little Hitcham, Jeremy Keen must establish both the capital cost and the value of benefits that will result. Jeremy decides to analyse the PV investment for a single house.

The new ZICER building for environmental research at the University of East Anglia (RMJM architects) incorporates photovoltaic cells on the glazing of a high level atrium space. The PV installation is a demonstration project and a research resource, as well saving money and conserving fossil fuels.

The capital cost has two parts. First, the PV cells and control gear plus the cost of installing them; and second the extra cost of building a house which is suitable for PV. The Little Hitcham houses have a floor area of about $100m^2$ on two storeys, so the roof area is $50m^2$; with a pitched roof, half the roof faces south, so a $25m^2$ array of PV cells can be installed. Currently PV cells cost about £600/m^2, so the PV cells for one house will cost about £15,000. However, two-thirds of this can be reclaimed through a government grant, so the cost to the Heathland Housing Association is £5,000.

The extra cost of building a house which is suitable for PV includes provision for mounting and maintaining the PV cells and cable ducts from the roof, and also extra costs that are incurred because PV-efficient houses must be laid out on the site so as to give south-facing roof pitches with minimum overshadowing. The preliminary estimate for these costs is about 2% of the typical £80,000 construction cost of a new house, or about £1,600. So the extra capital cost for installing PV is £6,600 per house.

Now the value of benefits. A typical family house uses about 6,000kWh of electricity per year, which is bought from the supply companies at a current price of about 6.5p per kWh: the cost of electricity is about £390 per year. PV cells generate about 100kWh per year per m^2, so the $25m^2$ array will generate 2,500kWh. This electricity would otherwise have to be bought from the supply companies, so the PV installation gives a cost saving of 2,500kWh x 6.5p/kWh = £162.50 per year.

Jeremy decides to do a standard whole-life costing evaluation of this investment, using a capital cost of £6,600, annual savings of £162.50, and a discount rate of 5%. This is the mortgage rate on the loan taken out by the Housing Association (7%), less the current rate of inflation (2%). Jeremy takes a 20-year period for the WLC study period, as this is the expected life of PV cells.

Using these assumptions the WLC study shows that the net present value of the PV installation is negative, because the value of the savings from Year 1 to 20, £2,025, is much less than the capital cost in Year 0, £6,600 (this is the cost to Heathland Housing Association, not the actual cost of the PV installation). On this basis, PV is not a good investment for the Little Hitcham project.

The figure for the present value of savings is almost certainly too high. Jeremy used the 5% discount rate that applies to the Housing Association development as a whole: it is a low rate, with a 2% risk premium over the 3% risk-free rate, reflecting the low risk attached to social housing investment; but an energy-saving investment requires a higher risk premium as future fuel prices are highly uncertain. Using a higher discount rate, say 8% per year, in a standard WLC calculation, the estimate for the present value of PV savings is reduced to £1,595.

In financial terms, Jeremy's first assessment tells him that the net benefit of the PV investment is negative – net benefit being the present value of the benefits less the costs (net benefit is another term for the net present value of the project cashflow).

Jeremy Keen is discouraged. He wants to check whether he has given PV a fair assessment.

Uncertainty about benefits

As is typical, the data for Jeremy's first assessment of PV at Little Hitcham includes a reasonably good estimate for the capital cost, but the estimate of future benefits is much less certain. The amount of electricity generated by 25m^2

of PV cells is the same whatever the price of electricity, but the financial value of the saving depends on the cost of the electricity that would otherwise have been bought from the supply companies. Jeremy's standard WLC analysis assumes that the future cost of electricity will remain the same as today's in real terms, although this is inconceivable. The price of electricity is sure to go up or down – probably both – over a 20-year period.

A simple and effective way of mapping future electricity prices is with a binomial tree, starting with today's price and covering the range of possible future outcomes. The 'spread' of the binomial tree can be calculated from past volatility, or it can represent the decision-maker's best estimate of uncertainty. The latter may seem unsatisfactory, but the person who makes the decision on photovoltaics has of necessity to work within the framework of available knowledge, even if it is no more than gut feel. In fact, historic fuel prices are one of the few good data sources relating to construction which allow the calculation of past volatility. Using this data, Jeremy estimates that the price of electricity could rise or fall by as much as 27% in a year. Importantly, he also believes that electricity prices will rise gradually in real terms; this gut feel doesn't translate easily into numbers, but Jeremy decides that at each node in his binomial tree there is a 55% chance that the 'rising' path will be taken, and a 45% chance that the 'falling' path will be taken.

Using these assumptions Jeremy plots his binomial tree for future electricity prices. With the data on this tree, the present value of expected savings from a PV installation comes to £2,560, taking account of price volatility, a higher discount rate for energy-saving investment, and the probability of a gradual electricity price rise in real terms. This is higher than his first estimate, but the message that a PV installation is not cost effective today is confirmed.

The situation is more interesting because the capital cost of PV installations is likely to fall over time. Everyone assumes that PV will get cheaper and more efficient. This is what has happened with other new technologies, like computers and mobile phones. Jeremy can reasonably

Binomial trees for a photovoltaic installation (1)

technique 11.1

The first binomial tree is for electricity prices. Starting at today's price of 6.5p/kWh, it can rise or fall by 27% in a year. At each branch:

$$e_u = e_i \times (1 + w) \qquad e_d = \frac{e_i}{(1 + w)}$$

where: e_i is the price before the branch, e_u is the price which rises at the branch, e_d is the price which falls at the branch, w is the rate of change. Here, e_0 = 6.5p/kWh, w = 0.27.

The second binomial tree is for the value of annual saving from PV generation. At each node:

$$S_i = e_i \times K$$

where: S_i is the value of electricity saved, e_i is the price of purchased electricity, K is the quantity of electricity generated by PV. Here, K = 2,500kWh.

The third binomial tree is for the benefit, or present value, of 20 years of PV savings, beginning with the value of savings specified for that node. Future savings are discounted back along the branches, according to the following formula:

$$B_i = S_i + p_u \times \frac{S_u}{(1 + r_e)} + p_d \times \frac{S_d}{(1 + r_e)}$$

where: B_i is the present value of savings at node i, S_i is the value of saving at node i, S_u is the value of savings on the rising branch, S_d is the value of savings on the falling branch, p_u is the probability of rising, p_d is the probability of falling, r_e is the discount rate for energy-related investments. Here, p_u = 0.55, p_d = 0.45, r_e = 0.08.

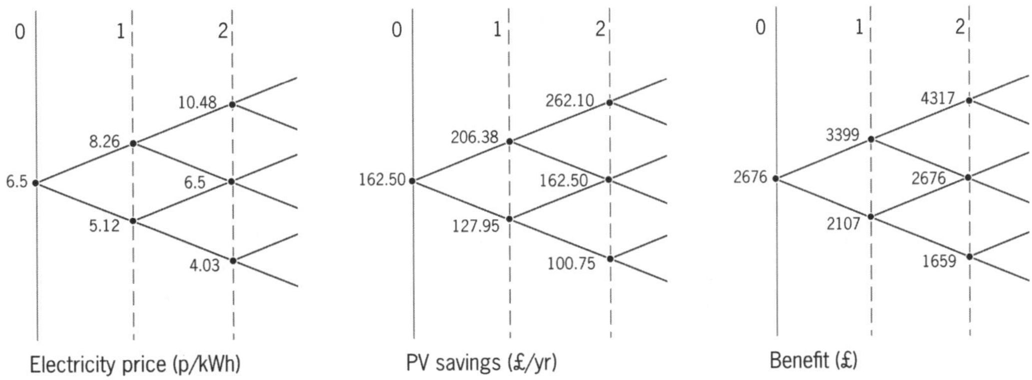

expect that in, say, five years a PV installation at Little Hitcham will cost less than it does today in real terms, and will probably be more efficient as well. It's hazardous trying to predict technical progress, but one UK government report estimated that PV generation of electricity would be twice as cost-effective in 10 years as it is now, equating to a decline in capital cost of 8% per year on average.

Jeremy can now think about whether a PV installation would be good value next year when it will be cheaper than today, or in later years when it will be cheaper still. It is clear that if the cost of PV drops as is expected, and the price of electricity rises as is perfectly possible, there must eventually come a point where the benefits outweigh the costs and an investment in PV would give a positive net benefit. At that point the Heathland Housing Association would want to sign up.

The option to install PV in the future

Jeremy's analysis shows that it is not good value to install PV today, but it is quite possible that it will turn out to be good value at some time in the future. So Jeremy wants the lifecycle option to install PV in the future. By acquiring the option to install PV, rather than installing PV now, he avoids a bad investment today but doesn't lose the chance to make a good investment in the future.

To install PV later, the houses at Little Hitcham must be built with south-facing roofs and facilities for a future PV installation, at a cost of £1,600 per house. This is the cost of acquiring the option to install PV. Note that it would also bring some other benefits, for example the option to install solar water heaters, whose value would be additional to the PV option. But what is the value of the PV option?

The estimate involves a number of steps, beginning with the binomial tree of future electricity prices. A new binomial tree of annual savings is created, in which the value at each node is the electricity price multiplied by the amount of PV electricity generated per year (which is the same for all nodes).

Even though it may not be cost-effective to install PV at today's prices, careful consideration should be given to building new houses with the option to install PV at a later date.

Binomial trees for a photovoltaic installation (2)

technique 11.2

The fourth binomial tree is for the net benefit of installing PV, that is the saving less the capital cost of installation. At each node:

$$N_i = B_i - C_i$$

where N_i is the net benefit, B_i is the benefit, C_i is the cost of PV installation. Here, PV begins at £5,000 in Year 0 and falls by 8% each year. The diagram shows the upper part of the binomial tree at Year 5, where there are the first positive values for net benefit.

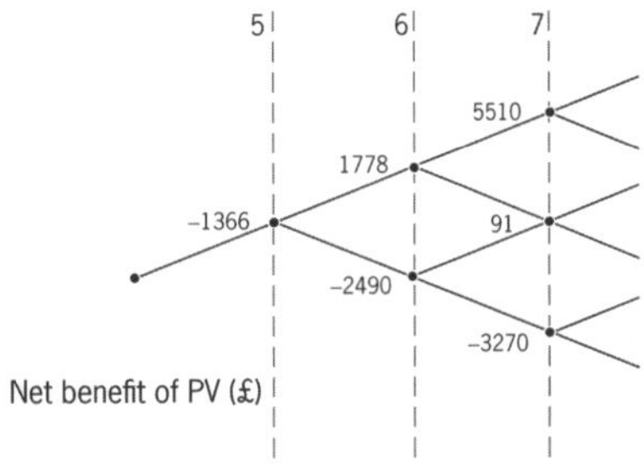

The fifth binomial tree is for the value of installing PV (V_i), that is the net benefit if it is positive, or 0 if the net benefit is negative, $V_i = \text{MAX} \{ N_i , 0 \}$

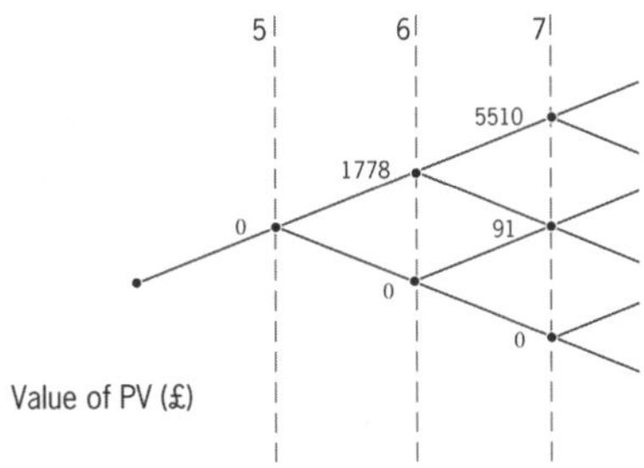

The option value is calculated from the value tree. At each node the value of installing is compared to the weighed average of the up and down values of installing PV in the following year. If the weighted average is higher than the value of installing, it is better to hold the option. The calculation is carried out at each node successively, beginning at Year 19:

$$Q_i = \text{MAX} \{ V_i , p_u \times \frac{Q_u}{(1 + r_q)} + p_d \times \frac{Q_d}{(1 + r_q)} \}$$

where Q_i is option value at node i, V_i is value of installing PV at node i, Q_u is option value on the rising branch, Q_d is option value on the falling branch, p_u is the probability of rising, p_d is the probability of falling, $p_u + p_d = 1$, r_q is the discount rate for option value. Here, $p_u = 0.55$, $p_d = 0.45$, $r_q = 0.05$.

This leads to today's option value in Year 0, Q.

The next step is to create a binomial tree where each node has the benefit (or present value) of 20 years of PV savings, if PV was installed at the corresponding electricity price. The present value is the weighted average of the savings in 20 subsequent years, weighted for probability and discounted at 8% per year.

A further step is a binomial tree of net benefit – which is the benefit less the cost of PV installation. PV costs are assumed to fall year-by-year, and the appropriate price level is used to establish the net benefit at each node. In the tree of net benefits there will negative values at many nodes where installation cost exceeds benefit, and

Net benefit, value and option value

principle 11.1

When new technologies are in their infancy they are often expensive and inefficient, compared to later, more developed versions – think of personal computers or mobile phones. For most users, PV is not cost-effective now and is unlikely to be in the immediate future, so in the early years the net benefit of the technology is negative. Only if electricity prices rise will PV generate a positive net benefit.

When the net benefit is negative investors do not buy the new technology, so its value is zero. Only in later years when its net benefit is positive do people buy it, and gain the positive value. The tree of values has zero and positive values, but no negative values.

Investors do not buy in the early years, but they want the option to benefit from the new technology if and when its value becomes positive. The amount they would be willing to invest in this option is equal to the present value of the positive segment on the tree of values, weighted for probability and discounted to Year 0.

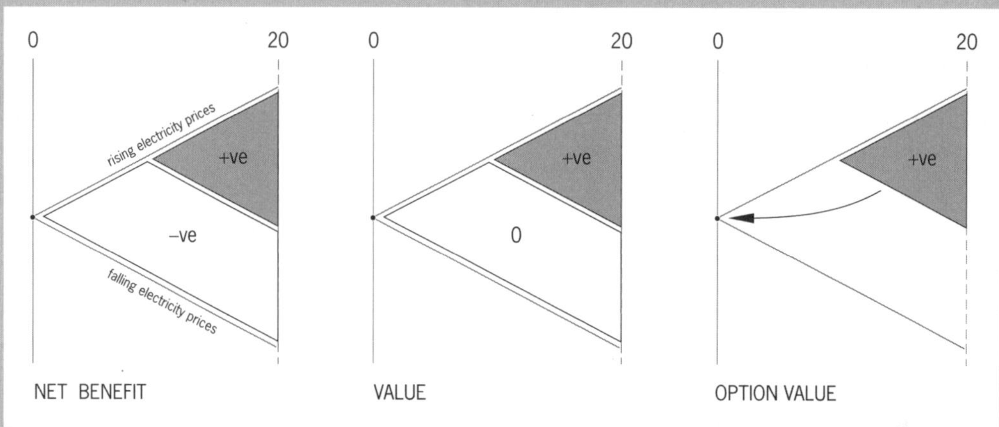

positives at the nodes where the price of electricity has risen and the cost of PV has fallen.

The final step is a binomial tree of the value of PV. Where there is a negative entry in the tree of net benefits, there is a zero in the tree of value: this is because the investment will not be made and no loss will be incurred – zero benefits will accrue. Over the whole binomial tree of values there are nodes with zero values and nodes with positive values, but no negatives: the aggregate of all these values must be positive.

The value of the option to install PV is the result of discounting all the positive entries on the value tree to Year 0. For this exercise a low discount rate is appropriate, say 5%, as there is little risk attached to the option: it will only be exercised when it produces a benefit, so there is minimal risk of a loss.

The value of the option to install PV turns out to be £1,832, slightly more than the £1,600 additional cost of building PV-enabled houses. This is encouraging – even though a PV installation is out of reach, it seems that Jeremy will be able to build PV-enabled houses at Little Hitcham.

The conclusion is based on the input data for the evaluation model. There are some important numbers:

- *discount rates*: there are two discount rates, a higher one for calculating the present value of future savings if PV is installed, and a lower one for calculating today's option value due to possible future savings: lower values for both discount rates increase option value
- *volatility of future electricity prices:* high volatility increases option value – if there is no volatility there is no option value
- *rate of rise in real electricity prices*: rapid rise increases option value
- *rate of decline in PV installation costs*: rapid decline increases option value.

There are other sources of uncertainty which could impact on the option value, including the future level of government

The top of this parking meter incorporates a photovoltaic cell, even though it is located in the City of London, the most dense urban area in the UK. The power demand is small and it is more convenient to use PV than lay an underground power cable. And parking is not regulated at night.

grants and the service life of PV systems. Equally, the cost of acquiring the option may not be accurate: if the estimated 2% additional cost for making PV-enabled is too high or too low, it will affect the decision about whether to acquire the option.

Having built a model to estimate the value of the option to install PV, Jeremy must go back and check all his assumptions, and where the assumptions are based on gut feel he should see what happens if alternative values are used – this is sensitivity testing. Jeremy finds that quite small changes in his assumptions can cause significant changes in option value. It is not reasonable

Financial appraisal of sustainable investment

principle 11.2

The case study example in this chapter concerned the generation of electricity in the UK with photovoltaic cells mounted on buildings. The example was imaginary but the conclusion is correct for almost all building in the UK: today the value of electricity generated is insufficient to make PV cost-effective. It is only cost-effective for small electricity demands where there is no mains electricity supply, for example, boats, isolated buildings, weather stations, parking meters.

So why do people install PV cells on UK buildings? There may be other benefits in addition to the value of electricity generated: educational, image-building/prestige, capability-building, research. Investors could take the rational decision that benefits like these, in addition to the electricity generated, would justify the cost of an installation.

Alternatively, it might be said that financial appraisal is inappropriate for an investment that is concerned with long-term sustainability, not short-term profit. But some appraisal is necessary, to avoid incurring opportunity costs – the costs of not investing in ways which would have produced greater sustainability benefits from the same resources. It is hard to imagine that any non-financial appraisal that compared the resources invested in PV with the sustainability benefits could demonstrate resource-efficient performance for today's PV systems. Financial appraisal is a convenient yardstick, and probably produces robust results.

It might be argued that many investors are overly risk-averse about renewable technologies, when the national interest would be better served by a more adventurous, risk-neutral approach – one in which the benefits on successful renewable installations would be balanced by the losses on disappointing ones. That is to say, the government might prefer investors to discount future savings from renewables at the public sector rate of 3.5% per year, which is virtually the risk-free rate, rather than adding a substantial risk premium. This may be why the UK government offers 65% subsidy for the capital costs of PV in housing, with the investor paying only 35% of the capital cost: the present value of a 20 year stream of savings from a PV installation at a highly risk-averse 20% per year discount rate is equal to 35% of the value of the same stream of savings at the risk-free discount rate of 3.5% per year. The subsidy would 'trick' a risk-averse investor into making an investment which is in reality cost-effective only at the risk-free rate.

to believe that a precise value for the option can be determined. However, Jeremy decides that for the new houses at Little Hitcham, it would be reasonable to spend about £1,500–2,000 per house to acquire the option to install PV later. This enables him to move ahead with project development.

Responses to change in technology

Technical progress is relentless but unpredictable. Many buildings are rendered obsolete when the function they were designed to accommodate is superseded – such as telephone exchanges for electro-magnetic switchgear, seaplane passenger terminals, ice-storage buildings, and factories for steam engines, slide rules or film cameras. This is natural and can hardly be considered wasteful, if the buildings served their purpose for a reasonable period of time.

There are cases where designers allow for the continued growth of current technology, only to be overtaken by technical change. Information technology in commercial buildings is a classic case. The expectation of continued growth in cabling led to ever deeper raised floors, but wireless technology means that the need for cabling is diminishing. The heat output of early personal computers led to predictions that the greater use of PCs would generate very high cooling loads, so designers provided increased air-conditioning capacity, but PCs have become more energy-efficient and so produce less heat.

Building in a way that takes account of the future development of technology would be an effective way of avoiding obsolescence due to technological change – if future technology could be anticipated. Attempts at prediction bring a severe risk of bad outcomes. At New Hall, a new women's college in Cambridge designed in 1960, the designers tried to anticipate future power supplies and concluded that electricity would increasingly be the dominant fuel – so designed an all-electric building (see p. 30). Their predictions were wrong, and their design makes it virtually impossible to retro-fit non-electric heating. Acceptance of uncertainty, and a design which incorporates the lifecycle option to change the heating

The PC and its networking do not sit easily in the offices of the earlier part of the twentieth century. The new technology may be powerful but it is also very demanding in environmental terms.

Francis Duffy *The Changing Workplace* (1992)

WLAN (wireless local area network) technology enables an enterprise to do all this [networking] without spending hard dollars on wiring, cabling, installation and construction. ... The cost savings that were recorded came from the few who actually did away with cabling, especially where this avoided difficult and expensive routing, such as in old buildings without raised floors. You cannot easily build a cabled network into a listed Georgian conversion.

British Council for Offices *Wireless technology and offices over the next five years* (2005)

system, would have been far more useful to the college than over-confident predictions.

There is a risk of over-investment in making provision for new technology that doesn't yet exist but is expected. At the University of Cambridge, where research laboratories are constantly having to introduce the latest technological developments, the estates offices work on the principle that they will never invest in provision for new technology that is expected to arrive more than three or four years in the future, because of the high risk that the provision will never actually be used, and the investment wasted.

On the other hand, buildings can be surprisingly resilient in the face of changing technology. Gloomy predictions of obsolescence are as fallible as any other kind of prediction.

Lifecycle options on new technology are much less risky than commitments to new technology, and can help to avoid under- or over-investment. In the case study of photovoltaics we saw that a modest investment in the lifecycle option to install PV could be cost effective. This is a realistic example: it is reported that the UK Building Regulations may be revised to require that new homes should be PV-enabled. And at the ZICER research building at the University of East Anglia, construction began before it was known whether a grant for the PV installation would be forthcoming. The building was designed with the option to add PV, and the option was exercised before the end of the construction period.

However, a lifecycle option is itself a commitment, and it is possible to under- or over-invest in options. It is important to be aware of technological uncertainty, identify lifecycle options that could help to manage the uncertainty, evaluate them as rigorously as possible, and only then decide whether it is most effective to incorporate new technology, do nothing – or acquire an option.

If a transmitter could be made weighing not more than two or three pounds with a range of 100 miles it would be useful to a great many people. ... Housewives should not anticipate their wholesale use for calling up the husband in his car and telling him to bring home a loaf of bread. The ether is already so cluttered up with a little bit of everything that a few million additional messages along towards dinner-time would prove fatal.

C C Furnas *The Next Hundred Years* (1936)

Kings Cross station was built in 1852 and survives despite enormous technological change. This would have amazed many Victorians:
I once heard a prominent civil engineer say that every railway station ought to be pulled down and rebuilt every ten years. The growth of traffic and of modern wants is rapid, and such structures are often built for utility alone.

E M Barry *Lectures on Architecture Delivered at the Royal Academy* (1881)

12

VERY LONG-TERM PROJECTS

Under conventional discounting, benefits that are received very far in the future have negligible present value, even with low discount rates. But there are some very long lasting buildings to which we do attach value. To model value in these circumstances a different approach to discounting has been proposed. However, the idea of very long-term evaluation remains problematic.

An authentic restoration

At first sight the old Militia Barracks in Milltown is a rather dull building. Built in a hurry in 1856, in response to an entirely bogus French invasion panic, it has survived in a surprisingly complete state – if you ignore decades of neglect and piecemeal alterations. A group of local enthusiasts want to convert the building into a Museum of Milltown Life, and have applied for a £6m heritage grant from the Lottery Committee. The application has landed on the desk of Harriet Smith who has to put together a project assessment report.

If the grant is approved, about half will be spent on refurbishing the buildings. James Oulton, a well-respected conservation architect, has proposed a meticulous restoration of the Barracks, undoing later accretions and reinstating the original fabric using authentic materials. This might seem straightforward as the 1856 construction was cheap and utilitarian, using off-the-shelf products of the day. However, the products have long since disappeared from builders merchants' shelves. In particular, the original architect, one Barclay Perkins, specified 'Fawcett's System' of fireproof floors, with ingenious permanent shuttering in the form of terracotta tubes which also acted as fresh air ducts. Various holes have been smashed

The old militia barracks in Grantham, Lincolnshire – a model for the fictional barracks in Milltown. Behind the pompous stone façade everything is utilitarian brickwork.

through these fireproof floors and James Oulton proposed to have new terracotta tubes specially made to repair the holes.

Even Harriet Smith realised that there are cheaper ways of repairing a floor. She decided to consult the Lottery Committee's in-house historic buildings specialist, Campbell Pritchard.

The value of finite resources

A back-of-an-envelope calculation showed that a brand new building for the Museum of Milltown Life would be cheaper than the authentic restoration of the barracks, as well as saving on running costs. The special floors weren't the only expensive proposal: other items included old-fashioned irregular glass, small Welsh slates, wooden laths and horsehair plaster.

The obsolete technology of Fawcett's System of fireproof floors caught Campbell Pritchard's imagination. It had apparently been quite popular in the nineteenth century, but he didn't know of any other surviving example. Over the years floors of this type become increasingly scarce: in effect, there is a finite, non-renewable resource of Fawcett's System floors, and if they are not cherished they might disappear entirely. Campbell Pritchard didn't like to think of that happening. He reflected that the prestige buildings of the Victorian age were being lavishly restored – St Pancras station, the Albert Memorial – but utilitarian buildings might slip into oblivion unnoticed. He supported the grant application, purpose-made terracotta tubes and all.

Harriet Smith received Campbell's opinion, scanned the business plan (it looked OK, as always), and wrote her report recommending approval. She looked forward to visiting Milltown for the first time – for the opening party.

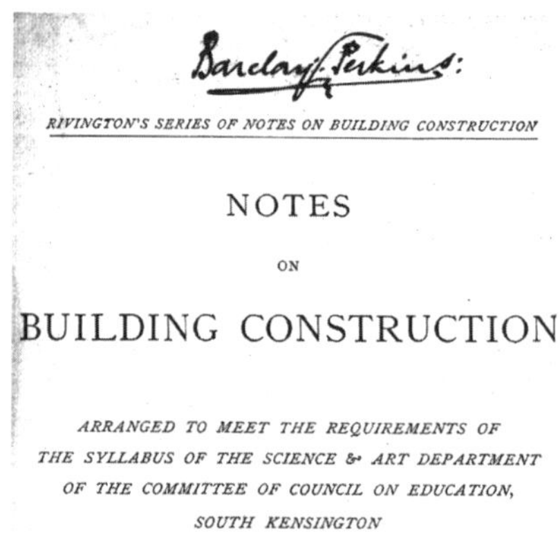
Barclay Perkins:

RIVINGTON'S SERIES OF NOTES ON BUILDING CONSTRUCTION

NOTES

ON

BUILDING CONSTRUCTION

ARRANGED TO MEET THE REQUIREMENTS OF THE SYLLABUS OF THE SCIENCE & ART DEPARTMENT OF THE COMMITTEE OF COUNCIL ON EDUCATION, SOUTH KENSINGTON

Barclay Perkins' copy of *Rivingtons' Notes on Building Construction*, which describes Fawcett's System of fireproof floors:

Fawcett's system consists of fireclay tubes, or, as they are called by the inventor, 'lintels', of the section shown at L, Fig.247, which rest upon the lower flanges of steel or iron I joists 2 feet apart. The lintels are placed obliquely between the joists – their own diagonals being at right angles to the joists. The spaces between and above them are then filled in with concrete – their lower sides are dovetailed as a key for plastering. ... This system has been used in numerous public buildings.

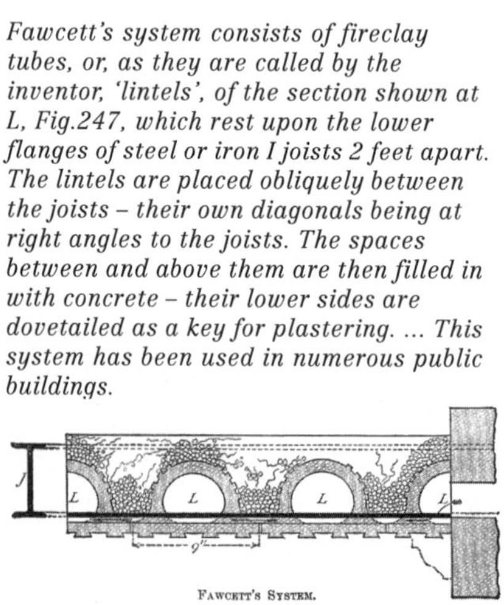

Fawcett's System.
Fig. 247.

Analysis of the very long term

What would Barclay Perkins, who designed the Militia Barracks in 1856, have made of the deliberations in 2005?

The Barracks were built in nine months and cost £3,600. Suppose that the cost was justified on the basis of 40 years use; we know that the mid-Victorian 'normal' rate of interest was about 3% per year, the figure given in the *Manual of Political Economy* by Henry Fawcett (no relation to the fireproof floor, or the author of this book). Inflation can be ignored as it was virtually nil. On these assumptions we can infer that the benefit generated by the Militia Barracks was worth £156 per year, because the present value of 40 annual payments of £156 at 3% discount rate equals £3,600 (at 1856 prices).

The Militia Barracks are still standing after 149 years. For some of those years they have been empty and neglected, but suppose they had generated a benefit of £156 per year continuously since 1856: the present value of 149 annual payments of £156 at 3% discount rate is equal to £5,128, in 1856 values; this is 42% more than was actually spent on building the barracks. Had such a

Refurbishment in progress in 2005 at the over-the-top Gothic Revival hotel at St Pancras station, London, which was designed by Sir Gilbert Scott and completed in 1874.

The option value of generic technology

principle 12.1

If you use generic technology it can be repaired or altered in the future by any competent builder; but if you use proprietary technology you have to go back to the original supplier, and if the original supplier has disappeared or changed their product range you have to use different technology or pay for specials to be made.

If Barclay Perkins had really worried about the long term, he wouldn't have used Fawcett's System of fireproof floors but heavier and more expensive brick vaults which could always be repaired or altered.

By paying more for brick vaults, he would have acquired the option to undertake repair or alteration in the future by any competent builder.

The choice between generic and proprietary technology still applies today. As the architect Peter Smithson observed: 'We had a '73 DS Safari and I had new cloth covers from France, they were the last in the world in that colour, and that's that – in eight years there are no more spares. If a building is built in the same way exactly the same thing happens, but at much shorter intervals: after two years you can get nothing, the extrusions are all unavailable, you can't buy the gaskets. ... The prefabrication theorists made an assumption about industry producing things as long runs, but in fact that's not the way it works' (interview by J Wojtowicz and W Fawcett, Hong Kong University, 1982).

long life been anticipated, Barclay Perkins might have specified more durable materials – something even better than Fawcett's System for the floors, for example.

If the Lottery Committee decides to fund the project, the Barracks will last a lot longer than 149 years – say, 250 years. The present value in 1856 of 250 annual payments of £154 at 3% discount rate is equal to £5,188 – only £60 more than the present value for 149 years! Barclay Perkins wouldn't have changed anything for £60, even at 1856 prices. Present values generated by discounting give negligable weight to very long-term benefits.

Investing for future generations

The Militia Barracks wasn't built for posterity, but the 175ft high Albert Memorial was. Designed by Sir Gilbert Scott and built in 1863-72, the monument to Queen Victoria's late husband cost £140,000, much of it raised by public subscription. The most expensive materials and the best workmanship were used. Was it meant to last for 100 years, or 250, or 500? Yes!

The Albert Memorial was a remarkable expression of Victorian taste, but rapidly went out of fashion. For most of the twentieth century it was despised. 'The Albert Memorial is the expression of purest philistinism' (Kenneth Clark, *The Gothic Revival*, 1928). 'The wonderful spark ... is just not there and no swing of the pendulum [of taste] can kindle it' (Ian Nairn, *Nairns's London*, 1966). It seemed that the Victorians' self-sacrifice was wasted on their ungrateful grandchildren; they might have been wiser to build a cheap, short-life memorial and spend the money saved on themselves.

In the 1980s the Albert Memorial started falling apart and was declared unsafe. In the 1990s, 120 years after it was finished, it finally came back into favour. £11m was spent on a comprehensive refurbishment, restoring the Memorial to its garish Victorian splendour. Its future seems assured.

But by some mysterious law of human taste it is almost impossible to enter into the spirit of the age that comes just before one's own. ... After a few years a period passes from shadows into the sunless impartial daylight of history ... That which has permanent value emerges, that which only appealed to a transitory phase of taste is finally obscured.

Lord David Cecil *Early Victorian Novelists* (1934)

The value of distant benefits

The Albert Memorial lasted for just about 100 years before it needed a major refurbishment. Suppose the Victorians thought that the £140,000 invested would be balanced by 100 years of pleasure or edification for the populace. At the Victorian 3% discount rate this implies £4,431 of enjoyment per year.

In fact, the Memorial did not produce very much enjoyment between 1902 and 1992. The present value in 1872 of 30 years of enjoyment (between 1872 and 1902) at £4,431 per year with 3% discount rate was £86,480 – embarrassingly less than the £140,000 invested.

If by investing a little more the Victorians could have extended the Memorial's life from 100 to 200 years with the same annual quantity of enjoyment, how much more might they have invested? Using the same discount rate, the answer is £7,285. And if the enjoyment could then have been extended from 200 years to 500 years, how much more? £399.

These figures defy common sense. If you care about the long term, as did the mourners for Prince Albert, three centuries of lugubrious communication with future generations has got to be worth more than £399, even at 1872 prices.

These unconvincing figures are calculated by discounting at 3% per year. This is quite a low discount rate, but in the very long term it still reduces future values to practically nothing. Discounting seems to produce unsatisfactory results for assets that have very long-term value. This would include not just the Albert Memorial but all artefacts and natural resources that we want to preserve for future generations.

The Albert Memorial in Hyde Park, London, 1863-72, after a faithful and expensive restoration in the 1990s.

The Albert Memorial is generally considered – and not unjustly – the perfect symbol of the High Victorian period. ... By the time it was completed in 1872 critics of advanced taste were already condemning it.

H R Hitchcock *Architecture: Nineteenth and Twentieth Centuries* (1958)

Time-varying discount rates

Economists have observed the problem with very long-term discounting and have come up with the theory that present values should be calculated with a discount rate that declines over time. This is based on the observation that people behave as if the difference of value between getting a benefit this year or next is larger than the difference between getting it in 10 years or 11, and much greater than the difference between getting it in 100 years or 101. The standard method of discounting assumes that the percentage drop in value is the same in all these cases; but in the time-varying approach, the percentage discount rate which is used for the near future gradually reduces for more distant benefits. Research suggests that a time-varying discount rate accords more closely with people's actual decision-making than the standard method of constant discounting.

The time-varying discount rate theory has now been adopted by the UK government. The 2003 revision of the Treasury's guidelines on public sector investment recommends a rate that starts at 3.5% per year but declines to 1% for benefits accruing beyond 300 years in the future.

We can investigate this new approach using the £11m Albert Memorial renovation as an example. To estimate the cashflow of benefits from the renovation, suppose that 100 years of benefits were anticipated: this equates to annual benefits of £400,000, because with standard discounting, the present value of 100 annual payments of £400,000 at 3.5% discount rate equals £11m. With this level of annual benefits, it is possible to compare the standard approach to discounting (constant discount rate) with the new time-varying approach. The results are shown in the table.

The time-varying discount rate increases the estimate of present value by about 4% for the 100-year duration, 8% for the 200-year duration, and 9% for the 500-year duration. These are not trivial increases, but they are certainly not radical.

Even with the time-varying approach, large increases in the duration of benefits generate disproportionately small increases in present value: an increase from 200 to 500 years increases present value by only £170,000, rather little for 300 years of benefits.

Duration	*Present value: 3.5% discount rate*	*Present value: time-varying discount rate*
100 years	£11,062,170	£11,499,155
200 years	£11,416,825	£12,278,386
500 years	£11,428,571	£12,447,732

Present value of Albert Memorial refurbishment for different durations and discount rates.

The mismatch between discounting and long-term value was apparent in the management of the great London estates, where substantial houses like those in Thurloe Square (above, by Basevi) were built on 99 year leases:

By actuarial tables the reversionary value of a lease with more than forty years left to run was said to be negligible. ... Yet landlords did not act as if it were negligible. ... The whole day-to-day business of an estate office would be unintelligible without the assumption that the first duty of the ground landlord was to pass on to succeeding generations the value of the property unimpaired and if possible enhanced.

D Olsen, *Town Planning in London* (1982)

The Treasury's time-varying discount rate does not seem to be very effective in mitigating the problem identified with standard discounting, that very long-term benefits have little impact on present values.

Changing values in society

Before completely dismissing the use of discounting for very long-term investments, it is worth looking again at the way the discount rate is established. An important component is 'rising wealth': assets that are highly valued now will be valued less in the future when society is wealthier. However, this process does not appear to be valid for the architectural heritage inherited from previous generations, so will probably not apply to the architectural heritage that we leave to future generations.

If the rising wealth assumption is invalid for architectural heritage, then perhaps a lower discount rate should be used, or even a negative discount rate, which would imply that benefits will be increasingly valued over time. We

Duration	*Present value: constant 1.5% discount rate*	*Present value: constant –1% discount rate*
100 years	£20,649,881	£69,062,170
200 years	£25,309,103	£158,552,747
500 years	£26,651,073	£6,047,832,469

Present value of Albert Memorial refurbishment for different durations with very low or negative discount rates.

Rising wealth hypothesis

principle 12.2

An important justification for discounting future benefits is the 'rising wealth' hypothesis. This says that society is expected to get wealthier, and because benefits received by wealthy people have relatively less value than the same benefits received by poorer people, future benefits received by a wealthier society will have less value than the same benefits received today. Hence, the value of a benefit is discounted if it is received in the future. Whenever discounting is used, it implies that the value of a benefit declines over time.

Rising wealth contributes 2% to the 3.5% discount rate recommended by the UK Treasury; the other 1.5% being 'pure time preference' (these are real rates which exclude inflationary growth).

The 'rising wealth' hypothesis seems to be valid for low quality goods that are superseded by higher quality goods (twin-tub washing machines no longer generate the enthusiasm they once did). But other assets become increasingly valued with wealth, for example, architectural heritage. Buildings that were regarded as worthless a generation ago and torn down for development would now be highly valued, protected from demolition and refurbished. In redevelopment projects in the UK it is now common to find old buildings retained and more recent buildings demolished.

The rising wealth hypothesis does not seem to be valid for architectural heritage, so discounting which is based on the rising wealth hypothesis is suspect.

can see what numbers this would produce for the Albert Memorial refurbishment, with £400,000 of benefits per year.

The results do not seem to make sense. The idea of spending £6bn on refurbishing the Albert Memorial is seriously implausible.

A fatal omission is risk. Realistic discount rates include a risk premium, and investing for benefits 500 years in the future is surely a risky proposition. We saw that the Victorians built a long-lasting monument that was despised after a generation, and similarly nothing built today could be sure to avoid falling victim to natural disasters, war or the cycle of fashion.

Perhaps discounting is the wrong approach for decision-making in very long-term projects – or perhaps the concept of very long-term decisions is questionable.

To secure stability in architecture a margin of strength is therefore required over and above the necessary demands of the moment. In actual professional practice this is not always easy to obtain. ... How many of our cathedrals would be left to us, if they had been built on a ninety-nine years' lease?

E M Barry *Lectures on Architecture Delivered at the Royal Academy* (1881)

The long term is lots of short terms

Is it realistic to think of making very long-term decisions? During the life of a building new decisions are constantly being made. Ely Cathedral is 900 years old, but today there is often scaffolding over some part of the fabric. The building's current state is the result of the actions and inactions by generations of bishops and surveyors. At Ely Cathedral each generation passes the building in a good state (hopefully) to the next generation, together with responsibility for its continuing care.

Designers who want to leave monuments for posterity should perhaps give more thought to the obligations they impose on their successors: the Albert Memorial was built with no expense spared and incorporated a complex wrought iron structural framework under the elaborate outer skin. But when the iron corroded, as was inevitably going to happen, there was no provision for renewal without taking the structure apart and rebuilding it again.

The reality is that every building which survives from the past comes with a multitude of options. The generation that inherits the building decides whether to retain, exercise or destroy the options. All a builder of monuments

The beautiful Japanese timber shrines at Ise date back 1500 years, but they are entirely renewed every 20 years.

The system of dedicating a new shrine every twenty years was conceived as a means of periodically renewing the fresh, pristine elegance of the shrine's sanctuaries. This ideal strives to preserve the vital beauty of the shrine for ever. ... This helps to explain why the oldest, most famous shrines preserve the fewest of their original structures. Other shrines abandoned this ideal and thus were forced to make repairs and other modifications.

Y Watanabe *Shinto Art: Ise and Izumo Shrines* (1974)

can do is pass on options which make it easier for his wished-for life history to be selected in preference to others. The option to maintain is very important, and the 'option to neglect' for the period from about 20 to 50 years after the building is completed is also advantageous – when buildings are out of fashion, maintenance problems can tip the balance of future decision-makers in favour of demolition.

The 'relay race' model of building for posterity suggests that it never makes sense to carry out a whole-life costing exercise with a 500-year cashflow. There can be very few decisions for which a cashflow longer than 50 years is meaningful.

Attempting to build for hundreds or thousands of years is not rational investment but wishful thinking.

The Roman Catholic cathedral in Los Angeles (Rafael Moneo architect) was designed for a 500 year life, and this led to the inclusion of lifecycle options:

It is designed to accommodate future developments. Large areas of flexible seating allow for a range of liturgies around the altar and font, infrastructure is designed for new technology, and nine of the eleven chapels will be completed in the future as devotionals for the patron saints of the city's ethnic communities.

Nicholas Roberts 'Design as materials research: building a cathedral to last 500 years' *Architectural Research Quarterly* (2003)

Summary

There is a significant parallel between the comments in this chapter on very long-term decisions and those we made on standard whole-life costing in Chapter 3.

On the shorter timescale of a whole-life costing study, it is over-simplified to suppose that the people who manage and maintain a building will perpetually repeat the decisions that the designer made when the building was first built. New decisions will be made in the light of changing circumstances.

Similarly, we suggest that it is not credible to think that any designer of monuments can make very long-term decisions which determine the future history of what is built today. The future will pass into the hands of other people and they will make their own decisions.

The impossibility of controlling the future does not mean that we should not care about the future. On the contrary, it is highly desirable that everyone involved in construction and real estate should take a long-term view. In practical terms, the best thing to do is to provide lifecycle options that are likely to be of value to future decision-makers.

The old Divinity School in Cambridge of 1878-79, by Basil Champneys (born 1842), is part of the heritage of the child in the buggy (born 2005) as a result of maintenance and refurbishment by successive generations.

13

LIFECYCLE BUDGETING UNDER UNCERTAINTY

In order to make rational investment decisions on the basis of first cost and running costs, it is necessary to estimate running costs. This is difficult both because of uncertainties, including uncertainty about the service life of building components. This chapter uses simulation to study budgeting when service lives are uncertain.

A big problem for a small project

Fairview Mews is a successful 25 unit cooperative apartment building in a North American city, completed five years ago. Most of the occupants are older people who have sold nearby houses. Each unit pays a share of the operating costs and taxes, and contributes $50 per month to a reserve fund for the future replacement of major building components, even though such work should not be necessary for many years. The level of contributions was set on the basis of a telephone poll of what similar projects were doing. Recently, legislation has been passed requiring that cooperative apartment buildings prepare replacement reserve plans, recording building components at risk and estimating how long they will last and how much it will cost to replace them. The unit-holders at Firthwood Mews saw it as a good update on the state of their building and a check on the adequacy of their reserve fund.

Fairview Mews, a five-year-old cooperative apartment building with retired residents who worry whether they will be able to pay for repairs.

OTT Consulting Engineers were appointed to evaluate the building. They produced a comprehensive Replacement Reserve Plan which suggested that the replacement reserve will be insufficient. They recommend that contributions should be upped from $50 to $80 per month for each unit. The Fairview management board is, however, hesitant to act. They know there will be opposition to increasing the monthly charge, and as a group they have doubts about the budgeting. Yet again, gut feel can be seen clashing with numerical analysis.

Understanding the life of building components

Before trying to quantify the problem it is important to think about the issues. Careful consideration of how buildings are maintained, and when and why components are replaced, may be the most productive stage of an analysis.

Sustainability

In many ways the Fairview Mews case is an instance of sustainability. The key idea of sustainability is that each generation must make provision for all succeeding generations, and in particular for the immediately succeeding generation.

At Fairview Mews, the enjoyment of the building by the current generation is accompanied by the gradual degeneration of building elements, such as roofs and mechanical systems. In order to pass on the building for the next generation's enjoyment, the present generation must make good this degeneration. To do this the current occupants put away an amount of money, which pays for reinstatement or can be passed on to future occupants to compensate them for the diminished state of the building. Future occupants can spend the money on repairing the building, or on something else if they prefer.

When do components fail?

It is to be expected that a given building component will not always have the same service life before failure, due to variation in such factors as climatic exposure, standards of maintenance, quality of original installation, and level of abuse, all of which influence longevity. Surprisingly, it is almost impossible to obtain data about time-to-failure distributions, although there are anecdotal hints about failure patterns: building managers will often express surprise that something failed so early in its life (although inevitably after the guarantee period), and also that something else lasted so long; these reactions define

Time-to-failure distributions

principle 13.1

Published data about the service life of building components is scarce and of poor quality. Amongst the few sources in the UK is the *HAPM Component Life Manual* which is written with the objective of insuring building elements against failure – if an element fails prematurely compared to service life, the insurance pool will provide compensation. A second source, the *BPG Building Fabric Component Life Manual*, is based on questionnaires sent to experienced building professionals asking how long various elements last. Their thoughts about typical, minimum and maximum life expectancies were collected. The ranges show the uncertainty about component lives, even among people who should know. For example, typical life expectancies estimated for PVC-U windows range from 11 to 100+ years, with a mean of 37 years.

The attempt to identify single point estimates of component life is not yet dead, but it is futile. It is built into the standard method of whole-life costing. Clearly the life of building components cannot be expressed in a single number. Lives are probabilistic and effort should be put into defining the best probability distributions.

The life-expectancy of any building component will follow some sort of distribution, and all distributions will have a mean. An important distinguishing factor that varies from one component to another is the 'peakiness' of the distribution about the mean. In a very flat distribution there is great variation in component life, from very short to very long. In a peaky distribution almost all components fail within a few years of the mean value. Just about all that can be gleaned from available sources is a rough estimate of the mean service life and the peakiness of the distribution.

Since available data is so poor, there is no possibility of achieving a close fit with any mathematically defined probability distribution; so it is reasonable to start by choosing a simple distribution for constructing a probabilistic model of service life. The lognormal distribution is a good choice: it is often used in reliability engineering; it can never have a negative value; very small values are extremely improbable; very high values are unusual but possible.

For various components the *BPG Building Fabric Component Life Manual* provides data which can be used for calibrating lognormal distributions – setting the mean and peakiness. This is the basis for a first attempt at service life modelling. It is far better than single point estimates, but far short of what could be desired. Research into service lives is most important – researchers should look for data to calibrate probability distributions. One difficulty is that by the time many building components reach the end of their lives, they are no longer in production, so there cannot be real service life experience of current products.

an approximate 'expected service life'. Although service life is uncertain, there is an unfortunate tendency to rely on rough estimates which somehow acquire the status of precise predictions.

In the Fairview Mews study, OTT Consulting Engineers assumed that some items, such as minor mechanical components, would be replaced a few per year, suggesting some sort of random failure process. For the major components they predicted the service life before failure;

although their figures were reasonable, there is a significant chance that component replacement would actually occur either earlier or later than the predictions.

When does replacement occur?

The when and why of building component replacement is not a simple matter. Some building elements, such as small mechanical components, do fail outright; but often failure is not so obvious.

It is almost self-evident to say that components are replaced when they no longer meet required performance standards. However, performance standards in buildings are rarely specified in a clear and explicit way, and they often change over time. The standards expected of a new first-rate office building may not apply in 30 years time, when the building is no longer prime space and attracts tenants whose priority may be low rents. A motivation for replacement is often the desire for higher performance standards, responding to rising expectations and technical advances, as when, for example, draughty but still functional single-glazed windows are replaced by better made double-glazed windows.

The time at which replacement becomes preferable to repair depends on many factors. Replacement can often be deferred, sometimes beyond the threshold when it would be cost-effective to replace. Perhaps the bottom line was revealed by a building owner who said that she would replace a defective roof 'when I have the money', and meanwhile would carry on paying for annual patching.

Another building manager, running a 12-storey apartment building, was faced with a cracked heat exchanger in an ageing roof-top air-handling unit. The mechanical contractor offered three viable alternatives: the crack could be welded and might last a few years; the heat exchanger could be replaced; or the entire air-handling unit could be replaced. The decision was dictated by the availability of money: the heat exchanger was replaced but not the whole unit.

A heavily patched, 30-year-old ashpalt roof that has been leaking for 10 years. When it is finally replaced more insulation will be incorporated and an elastic membrane will be used instead of asphalt.

A maintenance manager with a large portfolio of buildings said that the main reason for replacing major mechanical components was not their age but the non-availability of spare parts.

In these examples functional performance is the primary consideration, but wall and floor finishes are often replaced before they are worn out, because they are out of fashion – what could be called 'premature refurbishment'. Some styles and building types are more exposed to changes in fashion than others. If a building has moved down-market, replacement of finishes may become less important. A lesson for specifiers: focus fashion-conscious design on short-life components rather than long-life components – paint and carpet rather than glazed tiles.

The dictates of corporate policy can also cause premature refurbishment. An architect ruefully recounted the story of a prestige bank interior that was beautifully and expensively fitted out with counters, tables and other furnishings in fine dark woods, reflecting the solidity and worthiness of the bank. The interior was capable of lasting decades, but within five years the bank re-branded itself and ripped everything out, replaced with chrome and marble to reflect a more modern image.

It is evident that much replacement in buildings is discretionary, albeit with different degrees of urgency. A leaking roof or a non-functional heating system is more urgent than deteriorating flooring in a service corridor. Very few building failures are immediately and obviously life-threatening.

Risks relative to replacement reserves

The concern which caused legislation was the perceived risk that replacement funds might not have enough money for necessary repairs. That would mean that future occupants might hit a funding crisis, or be faced with sub-standard housing conditions.

However, there is another risk, that the reserve is over-funded, when the people now living in the building unnecessarily forfeit some degree of well-being because of excessive contributions to the replacement fund. Future occupants would gain a windfall benefit from the over-

A panel of late Victorian tiles revealed when a 1950s fireplace was ripped out of an early Victorian house. They are now back in fashion and will be recycled. Glazed ceramics are one of the most durable materials – witness the collections in archaeological museums.

large replacement fund: they might be able to pay themselves dividends or cancel their own contributions, which is not the objective of sustainability.

Technical progress is one reason why component replacement budgets may be over-estimated. A manager of a 20-year-old building identified some examples: when replacing wood parquet flooring they now use an engineered wood flooring which takes 35% of the time to install; and the in-unit electric water heaters are now replaced with smaller, more efficient units. The new components are substantially cheaper in real terms, so a replacement fund that had been budgeted on the basis of the old technology would have collected too much money.

In some ways, the requirement for a replacement reserve is a book-keeping matter only. Minor work, such as the renewal of finishes (for example, paint, carpet, tiling) or minor components (for example, sealed double-glazed units, plumbing), is carried out by the building staff and charged to the annual operating budget, so never appears as replacement expenditure chargeable to the reserve.

How should a group such as the Fairview Mews residents deal with all of this?

Analysis of component replacement

Norman Sirkin, a retired airline executive living at Fairview Mews, has time on his hands and decides to try some analysis. First of all he wants to identify the fundamentals of the problem.

It is possible that half of the items scheduled on the Replacement Reserve Plan of OTT Consulting Engineers are likely to be covered by routine operating budgets. These are cyclical renewals, small elements in larger systems, or small items relative to the project's overall operating. If one or two smoke detectors are replaced per year, it does not make much sense to book the work to the replacement reserve fund – it is just part of normal maintenance. This instantly makes the analysis more managable. In Fairview Mews, as in most buildings,

Components are often replaced before the end of their functional service life, perhaps because of fashion, rising expectations, or because they are not worth re-using in a major refurbishment project.

relatively few high-cost items dominate the planning of component replacement.

Mr Sirkin decides to focus on eight high-cost components and carry out a simulation study, in which the time-to-failure of the eight components is the source of uncertainty. Simulation is particularly suitable for a system with many separate sources of uncertainty. Precise mathematical calculation of all the combinations and permutations would be extremely cumbersome.

The essence of the model is the way it deals with component life uncertainty. Mr Sirkin knows he is not looking for a single number for each component's service life, but a probability distribution. With scrappy data, he cannot even hope for precisely defined probability distributions. He decides to use lognormal distributions, which correspond to a relatively low probability of failure in the first few years of a component's service life, and a small probability of a very long service life. He calibrates the lognormals so that average service life matches the conventional expectation, as used, for example, in OTTs Replacement Reserve Plan. Given the poor data there is no point doing anything more complicated, but if new data showed that alternative probability distributions would give better results they could be used instead.

The simulation model generates a 50-year component replacement scenario, specifying the year (or years for shorter life components) in which each component is replaced. From this scenario a replacement cashflow is then derived. Every time the model is run it generates a new replacement scenario and the corresponding cashflow.

The model uses today's costs for future component replacement, in the absence of better data; that is to say, all costs are real costs, excluding inflationary price increases. It is reasonable to think that technical progress will reduce some costs in real terms, but rising expectations and perhaps new health and safety regulations may increase others. Thus the cost of component replacement in Mr Sirkin's simulation is the same as in the Replacement Reserve Plan; his simulation studies the

Maintenance planning is plagued by contradictory estimates for the service life of components. For example, there were doubts about Yorkshire stone paving slabs in the City of London:

After remaining down for 18 years in principal thoroughfares, the paving was removed and relaid in courts or in streets of inferior traffic, where, it was estimated it would last for 15 to 20 years longer. The total life was taken as 36 years.

But it has recently, in 1875, been estimated that the Yorkshire stone foot-pavements in London now last only 7 years in the busiest localities, or 12 years in localities of small traffic. There appears to be an error in the estimate.

D Kinnear Clark *Recent Practice in the Construction of Roads and Streets* (1890)

Simulation of component replacement cashflow

technique 13.1

The simulation generates a 50-year replacement history for eight major components at Firthwood Mews. The life of each component, or time to failure, is described by a probability distribution which follows a lognormal curve. Each component's lognormal is defined by two parameters, which approximately correspond to the average and standard deviation of the component life. The lognormal for the roofing system is shown below, together with the cumulative lognormal which is used with a random number for generating a simulated co

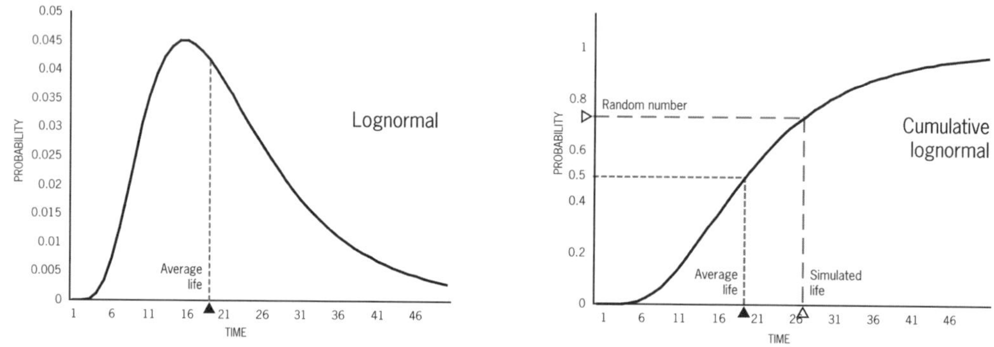

In the Firthwood Mews study the following components were modelled:

	Expected mean life (years)	*Standard deviation*	*Replacement cost*
Roofing system	20	4.5	$141,750
Elevator (lift) equipment	25	3.5	$142,800
Elevator (lift) interior	25	8	$63,750
Fan-coil mechanical units (25 units)	20	4	25 × $3,500
Exterior wall caulking	10	2	$15,188
Parking surfacing	25	3.5	$112,500
Parking sealing	7	1.5	$7,500
Entry security system	15	4.7	$22,500

The simulation begins with the first component. In Year 1 the component's lognormal distribution is used to generate a simulated service life. If the service life ends before Year 50, the process is repeated and a new service life is generated in the same way, and so on until Year 50 is reached. The same exercise is carried out for the remaining components. In the case of the fan-coil units in each flat, there are 25 separate simulations. This exercise results in a 50-year schedule of component replacement at Firthwood Mews.

Whenever a component is replaced, the replacement cost is added to the cashflow for the replacement year, so for each schedule of replacements there is a corresponding 50-year cashflow.

Every time the model is run a new schedule of replacements and the corresponding cashflow is generated. Using the model, 1000 runs were carried out in a Monte Carlo simulation.

way that the replacement costs might be distributed in time, not the magnitude of the costs.

Having set up the model, Mr Sirkin runs it as a Monte Carlo simulation with 1000 runs. This generates the average replacement cashflow, and also the probability of outcomes above and below the average. Time to talk to the Fairview Mews management board.

Managing the replacement cashflow

Mr Sirkin presents his findings to a special meeting of the Fairview Mews management board. The members are keenly interested to find out how much they will have to contribute to the reserve fund. Mr Sirkin explains that they start with two precise figures: $50 per unit per month as at present, and $80 as recommended by OTT Consulting Engineers. His own study does not result in a third precise figure, but reveals what is going on, so that the board can recommend a credible level of contribution.

The question of 'how much contribution is enough' is largely determined by the attitude to risk. If the residents want to be assured that they will never, ever have to borrow money or ask for a special contribution, they will have to build up an enormous reserve fund – much larger even than that recommended by OTT. There is a high opportunity cost of holding such a large fund, because the cash in the reserve is not available to the occupants for current consumption.

One baseline figure is the contribution that would meet the average cashflow requirement in Mr Sirkin's Monte Carlo simulation. This would be $70.53 per unit per month. Why not adopt this figure? The difficulty is that being an average of 1000 model runs, there are about 500 'unfavourable' runs in which cash is needed sooner, as well as 500 'favourable' runs in which the cashflow is less demanding than the average.

The 'average contribution' model should work in the long-run, but leaves the possibility of having deficits at some time during the study period. With the average contribution, Mr Sirkin's simulation suggests that there is about a 31% chance of the reserve experiencing a deficit at some point. This seems too high to the management board. They feel that a 20% chance of deficit, which means 80% chance of no deficit, is more comfortable. The deficits are temporary, when a replacement occurs before the reserve fund has grown sufficiently to pay for it, and are clawed back as further contributions are paid into the fund. In these situations a loan could almost certainly be arranged, or alternatively there may well be managerial discretion to defer replacement until funds come in. The risks do not seem too severe.

According to the simulation, a 20% chance of deficit requires a monthly contribution of $74 per unit. The board will take this figure to a vote of the residents. They are sorry that it is higher than the current 'guesstimate', but are pleased to have a good explanation for the increase, rather than having to accept an 'expert' assertion which was not convincing – and higher.

Before the meeting breaks up Mr Sirkin raises two important points. His simulation gives about as good a picture of the component replacement problem at Fairview Mews as is possible with current knowledge, but the building is still new and there is great uncertainty in the assumptions he used. Over time there will be new information about durability, cost, new technologies, new regulations, new expectations – the replacement reserve should be regularly reviewed in the light of better information.

Second, his analysis, like OTT's, has used real prices, but the prices that will ultimately be paid for component replacement will rise with inflation: the reserve fund must be invested to grow at least as fast as inflation, and the monthly contributions should be indexed to match construction price inflation.

Thinking ahead

It is interesting that the Fairview Mews board spent so much time debating the component replacement issue, when it was the lack of concern about funding replacements by condominium boards that led to legislation.

A possible reason why it was a big issue at Fairview is that the building is largely occupied by retired people. One might think that older people would have less reason for long-term planning than younger people, but a crucial factor is that older people cannot assume increasing wealth, and they are less confident about being able to cope with problems that may arise in the future. They are more willing to spend money now to avoid problems in the future. Older people apparently discount the future at a lower rate than younger people.

A condominium developer observed the same phenomenon in the design of new buildings, expressing some frustration at being unable to interest younger purchasers in energy-saving technologies, whereas these features were being incorporated in buildings for seniors.

Discount rates encapsulate an attitude to the future. A big factor in an individual's discount rate relates to the expectation of becoming wealthier or poorer in the future. People who expect rising wealth tend to discount the future more heavily: they are less willing to make sacrifices now for future benefits, because in the future, when they are better off, the benefits will have little value. Remember governments' routine expressions of anguish that young families do not save enough? People who expect to become poorer discount the future at a lower rate: they are much more likely to set aside present income for future benefits.

The attitude to sustainability exhibits the same pattern on a grander scale. People who are pessimistic about the future are willing to sacrifice present consumption for distant future benefits: they have a low discount rate. People who are more optimistic that environmental problems will somehow be solved see less reason to worry about tomorrow's problems: they have a higher discount rate.

Early American railroads reflected the values of a young and optimistic country:

American railroads were built at the lowest possible cost. While the British sought permanent right-of-way with broad curves and gentle grades, Americans paid scant attention to the curves and grades and accepted the fact that their railroads would require almost constant maintenance and replacement. Although British lines cost more money and took longer to complete, they were better built and lasted longer with less maintenance than American railroads. ... Although American railroads were not built to the standards of their British counterparts, they served the needs of a new nation with limited capital and labour. By the late nineteenth century most of the major lines of the country had been substantially rebuilt to eliminate sharp curves and steep grades.

Robert L Frey (ed) *Railroads in the Nineteenth Century* (1988)

Long-term budgeting

Many public sector buildings in the UK are now being built through public-private partnership agreements. One objective is to inject private sector competition and efficiency into public sector projects. Under such an agreement the building is constructed by a private sector consortium, who also undertake to maintain the fabric for a long period, often 30 years. This long-term commitment means that the consortium has to think about whole-life costing – the totality of first cost and running costs. They should be ideally placed to make trade-offs between the two and avoid under- or over-investment at the construction stage. This is a positive aspect of long-term agreements.

The public sector counterpart pays the consortium an annual charge for the duration of the agreement. Part of the annual payment is for building maintenance (it also repays the capital that the consortium borrowed to design and construct the building). Just like the residents at Fairview Mews, the consortium has to reconcile a regular annual payment with an uncertain cashflow for maintenance and the replacement of building components. As they have to tender for the project in competition, the consortia are under pressure to set a low annual payment in order to win the project, whilst wanting to minimise the risk of making a loss if the maintenance cashflow turns out to be more expensive than expected.

This is a serious problem for a large and complex building and the consortia are worried about bad outcomes, so they invest considerable effort in predicting exactly when components will fail and exactly how much they will cost to replace. This is done even for low-cost, short-life components, so a mass of data is created. The consortia then set up a budget for the maintenance and replacement costs in each year of the agreement, which can last for 35 years or more, to the nearest penny.

The precision is almost completely spurious and the effort almost completely wasted.

Maintenance budgeting cannot escape uncertainty, and making ever more precise guesses is not a helpful response. It would be more effective to acknowledge the probabilistic nature of component service lives and the replacement cashflow, and analyse the range of possible outcomes and the associated risks. Businesses are used to investing in risky projects, if the risks are clearly understood.

The aversion to risk, and the almost child-like pretence that it can be wished away, is unfortunately found in other aspects of public-private partnership agreements. The public sector users are required to state exactly what their activity requirements will be for the duration of the agreement – this could be a 30-year agreement for a hospital. It can't be done. Writing down guesses doesn't help, and spending several years writing down very detailed guesses only increases the likelihood that the building will be obsolete when it opens.

Both public sector users and private sector consortia are desperately in need of strategies to manage the uncertainty that they unavoidably face – uncertainty about building use and uncertainty about building durability. The exploitation of lifecycle options presents an important opportunity.

14

SELF-ASSESSMENT MATRIX FOR WHOLE-LIFE EVALUATION

This chapter presents a non-quantified evaluation tool for use by a building owner or project team. It distinguishes between situations where investment for long-term benefits is likely to be appropriate, and others where uncertainty makes it more attractive to seek lifecycle options. The objective is to reduce the risk of under- or over-investment.

Present vs. future

Whole-life costing, like sustainability, is founded on the principle that today's decision-makers should take account of long-term impacts, and aim for decisions that produce the best outcomes over the long term. This is for reasons of both self-interest and social responsibility.

The principle requires that present day and future costs and benefits should be weighed against each other. In typical construction projects capital investment comes at the beginning with benefits following over many years: the temptation to economise on present day investment must be offset by considering the value of future benefits. But it is not easy to compare costs and benefits that occur at different times.

In broad terms investors must take a view on whether, in a particular project, the expectation of future benefits makes it a good idea to invest heavily now; or whether the future benefits are only sufficient to justify modest investment now.

An investment which is insufficient to take advantage of potential benefits is inefficient – it constitutes under-investment; whereas investing heavily for small future benefits is also inefficient – it constitutes over-investment. Both these cases are to be avoided whenever possible.

The self-assessment matrix helps to identify the risk of under- and over-investment in a particular project, and gives guidance on how to balance present day investment against future benefits.

It can be used to assess a whole project, or part of the project – or even a single building component.

How much uncertainty?

If everything could be known about the future of a building, it would be easy to invest just the right amount in its construction and calculate which form of structure, which windows, which finishes, etc, would give optimum value over the building's life.

However, when there is uncertainty it is much harder to get the right balance and avoid under- or over-investment. This is where the risk-adjusted discount rate plays such a crucial role in whole-life costing calculations. High uncertainty leads to a high discount rate, which in turn leads to short-term planning.

The discount rate has a slightly mystical status, but it is no more than an encapsulation of the basic characteristics of the project and its environment.

The self-assessment matrix by-passes the discount rate. It goes back to the basic characteristics of the project, and uses them to indicate reasonable strategies for long-term decision-making.

There are five factors which have to be considered:

- restriction of capital
- uncertainty about technology
- uncertainty about regulations
- uncertainty about use
- restriction of time horizon.

If a project is highly restricted or uncertain in relation to any or all of these factors, it is harder to justify investing for the long term.

Assessing the five factors

To use the self-assessment matrix the project has to be assessed against five factors. This is done on a four-point scale, using first-hand knowledge of the project.

Restriction of capital

Very low	Organisation has ready access to capital – able to maximise long term value
Low	Investment considers both capital and running costs
High	Reducing capital costs is more important than reducing whole-life costs
Very high	Little capital is available for the project

- Consider whether there is adequate capital available, and whether capital and running costs be traded off against each other to achieve the best overall value.
- Decide which of the four categories best describes the restriction on capital for the project.

Uncertainty about technology

Very low	Mature and stable technologies only
Low	Technologies subject to gradual evolution
High	Possibility of dramatic shift in technology
Very high	Technologies that are likely to change rapidly and unpredictably

- Consider whether the project involves emergent or obsolescent technologies that are likely to change, or technologies that are mature and stable.
- Decide which of the four categories best describes the technology in the project.

Uncertainty about regulations

Very low	The regulatory regime will remain stable
Low	Only limited change in regulations expected
High	There could be major shifts in regulations
Very high	Radically new regulations are likely to be imposed

- Consider whether the project is subject to regulations that may change significantly during its service life.
- Decide which of the four categories best describes the regulations affecting the project.

Uncertainty about use

Very low	The use pattern is unlikely to change
Low	The use is subject to gradual, evolutionary change
High	There may be sudden, dramatic change in the use pattern
Very high	Sensitive to rapid, unpredictable change, eg fashion trends

- Consider whether the use of the project is expected to remain stable for the foreseeable future, or whether it is likely change in a volatile way.
- Decide which of the four categories best describes the expected use of the project.

Restriction of time horizon

Very low	Expected to remain an asset for an indefinite period
Low	Expected to remain in use for considerable time
High	Expected life is uncertain but may be short
Very high	Known to have a short life only

- Consider whether the project is a long-term asset, or for a short-term requirement; or whether the timescale is uncertain.
- Decide which of the four categories best describes the time horizon for the project.

Completing the self-assessment matrix

principle 14.1

Enter marks in the project self-assessment matrix, in the cells corresponding to the assessment of the five factors.

In this example, it has been judged that the project is rated 'low' for restrictions on capital, 'high' for uncertainty about technology, 'high' for uncertainty about regulations, 'low' for uncertainty about use, and 'very low' for uncertainty about time horizon. The 'pattern' of marks is shown on the bottom right.

1

	uncertainty			
	Very low	Low	High	Very high
Capital		*		
Technology				
Regulations				
Use				
Time				

2

	uncertainty			
	Very low	Low	High	Very high
Capital		*		
Technology			*	
Regulations				
Use				
Time				

3

	Very low	Low	High	Very high
Capital		*		
Technology			*	
Regulations			*	
Use				
Time				

4

	Very low	Low	High	Very high
Capital		*		
Technology			*	
Regulations			*	
Use		*		
Time				

5

	Very low	Low	High	Very high
Capital		*		
Technology			*	
Regulations			*	
Use		*		
Time	*			

6

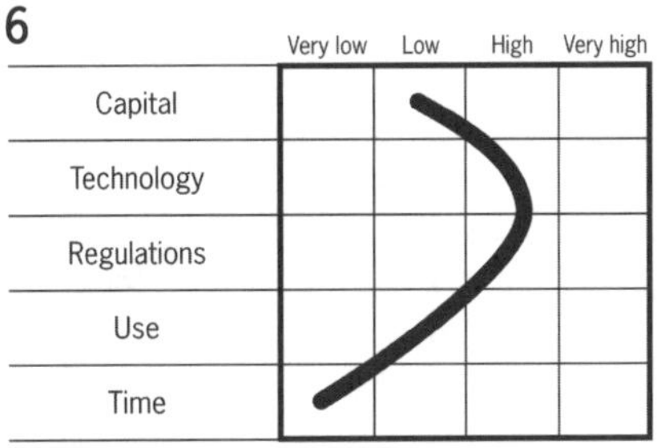

Comparison with templates

To assist with drawing inferences from the self-assessment matrix, the pattern of stars for a particular project can be compared to a number of templates.

Low uncertainty

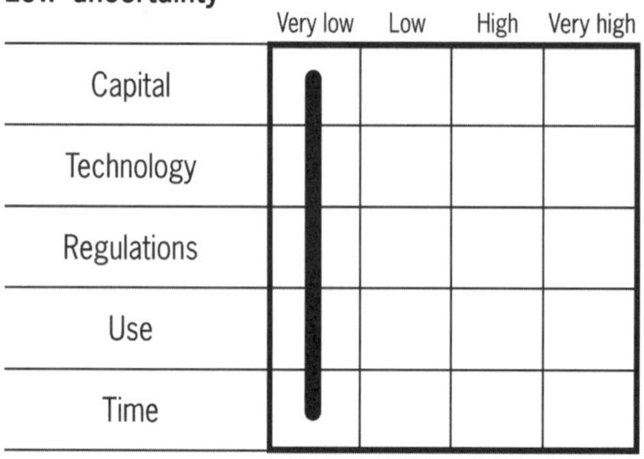

- The investment is 'safe'. Investing for long-term benefits should be considered.
- Traditional WLC methods are likely to be of use in identifying optimum investment and design decisions.
- Example: building for an Oxbridge college.

High uncertainty

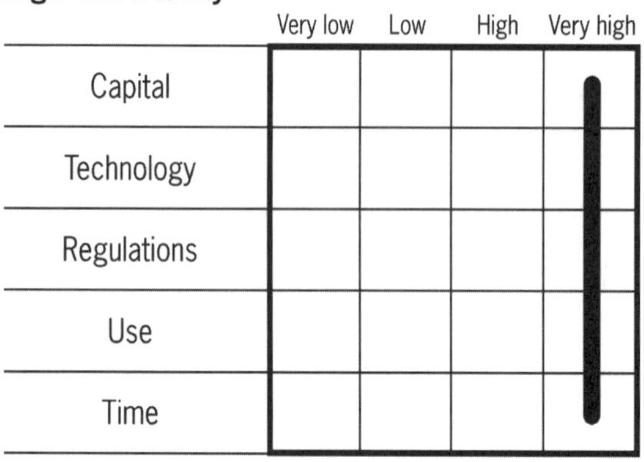

- It is not possible or sensible to invest for long-term benefits.
- It is desirable to incorporate any lifecycle options that can be afforded, to facilitate future change.
- Example: restaurant or fashion boutique.

'Convex' shape

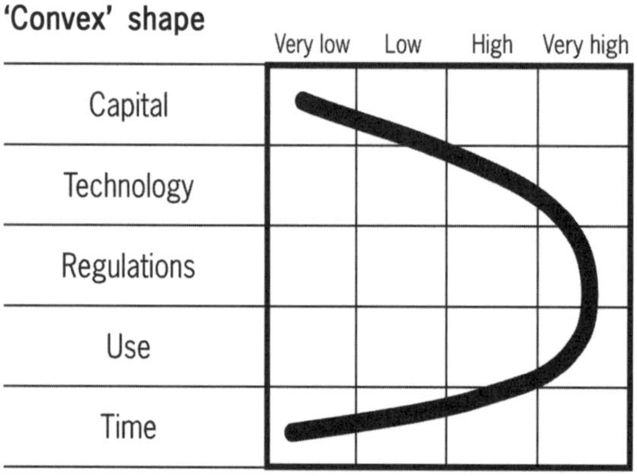

- Stable context undermined by uncertainty about technology, regulations and use.
- Risk of over-investment.
- Invest in lifecycle options to switch use, reconfigure or refurbish.
- Example: laboratories for high status university.

'Falling' shape

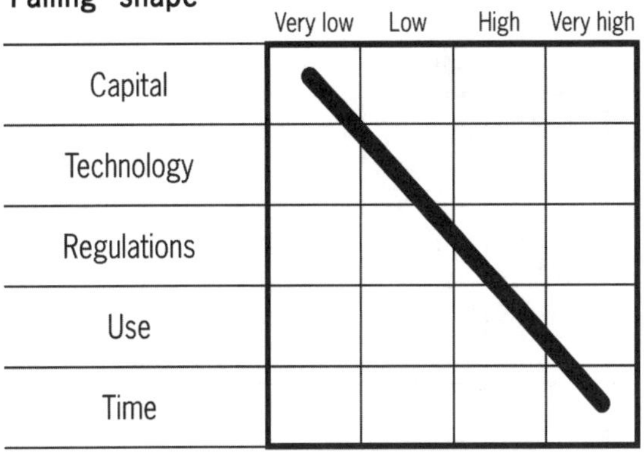

- Abundant capital, varying uncertainty but short time horizon.
- High risk of over-investment,
- Invest in lifecycle options to switch use, reconfigure or refurbish.
- Example: prestige buildings for temporary use.

'Concave' shape

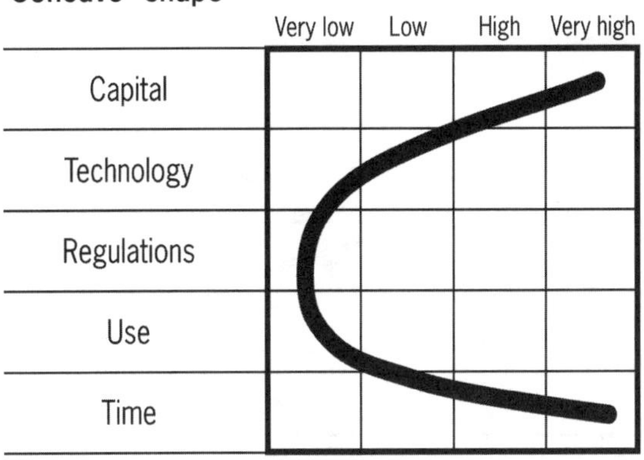

- Scarce capital and short time horizon, but low uncertainty about technology, regulations and use.
- Risk of under-investment.
- Incorporate lifecycle options to expand or refurbish, especially if they have no cost or low cost.
- Example: owner-occupied houses.

'Rising' shape

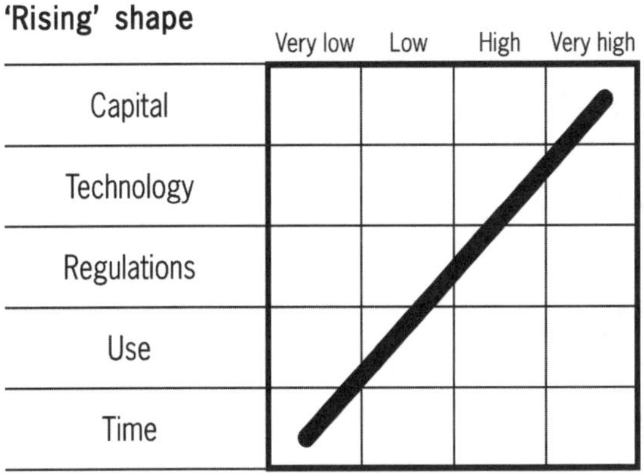

- Scarce capital, varying uncertainty, coupled with long time horizon.
- Very high risk of under-investment.
- Incorporate lifecycle options to expand or refurbish, especially if they have no cost or low cost.
- Example: primary school in overspent local authority.

Self-assessment form

technique 14.1

This blank self-assessment form can be photocopied and used for the evaluation of current projects or decisions. Describe the project or decision by marking the appropriate boxes, and compare the pattern with the templates.

	Very low	Low	High	Very high
Restriction of capital				
Uncertainty about technology				
Uncertainty about regulations				
Uncertainty about use				
Restriction of time horizon				

Low uncertainty

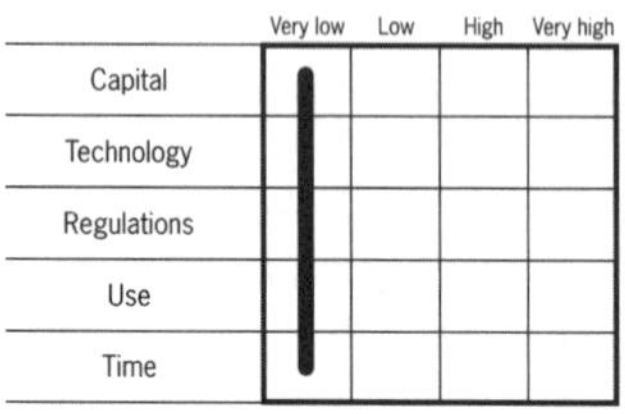

- Make long-term investments

High uncertainty

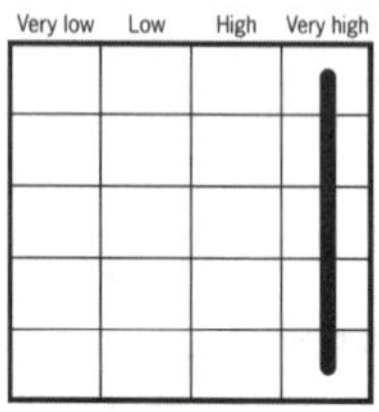

- Wait and see
- Seek low-cost lifecycle options

'Convex' shape

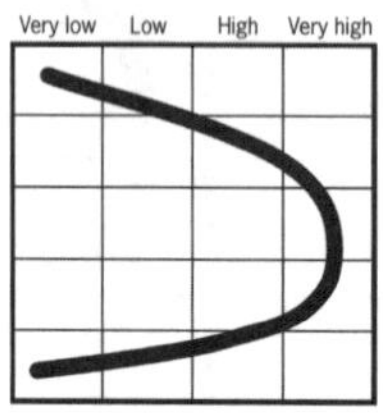

- Risk of over-investment
- Buy lifecycle options

'Falling' shape

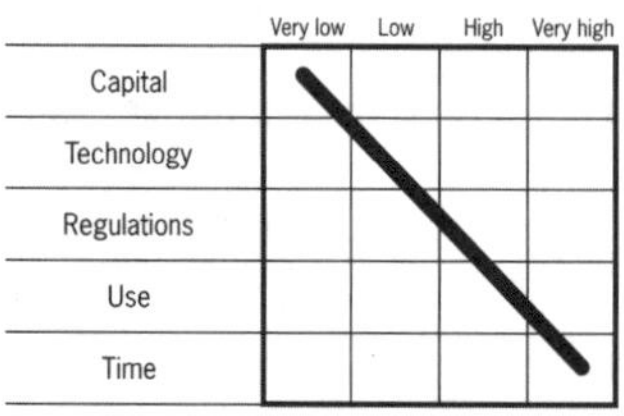

- High risk of over-investment
- Buy many lifecycle options

'Concave' shape

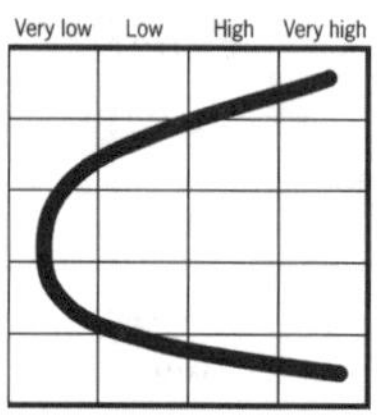

- Risk of under-investment
- Options are low priority

'Rising' shape

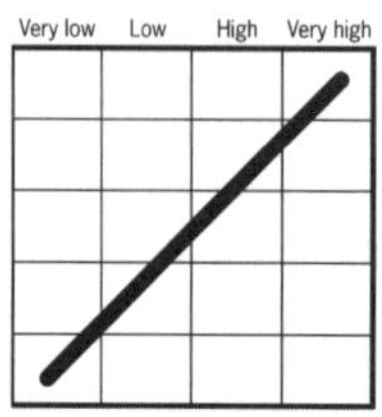

- High risk of under-investment
- Options are low priority

Follow-up activities

The self-assessment matrix aims to counter the perception that whole-life costing is so complicated that it has to be left to 'experts'. There is a danger that experts who are not closely involved in a project may apply complex and possibly inappropriate mathematical techniques, which produce answers that are baffling to the project team. Far better that everyone involved in decision-making has a good understanding of the basic principles of whole-life value. Then strategies for long-term value can be debated and tested, leading to a sound strategy that has the confidence of everyone who has to implement it.

It raises awareness of critical issues and de-mystifies whole-life costing, which sometimes seems over-technical and detached from experience.

The self-assessment matrix is a simple tool that helps non-specialists get started with an investigation of whole-life costing issues. Ideally it should increase the motivation for investigating the issues more deeply by going on to use quantified methods.

Conclusion

The analysis of economic behaviour and investment decisions continues to evolve. People traded goods and invested in their future long before the discipline of economics, let alone business finance, was invented.

The advance of knowledge should lead to better decisions, but in the examples in this book we have shown several instances where the gut feel of experienced practitioners produced better decisions than the application of standard whole-life costing models. However, experienced practitioners are a scarce resource, which cannot be replicated cheaply or quickly.

Experienced practitioners make better decisions because they analyse problems in a more effective way, even if they sometimes find it hard to articulate how they do it. They present a challenge: the development of better models with equivalent decision-making abilities. We hope that new generation whole-life costing is a step in this direction.

We expect that some people may have doubts about new generation whole-life costing, but we feel that everyone should share our objective of improving decision-making for construction and property investment, so that it is based on a rational, long-term view of the future.

WHERE TO FIND OUT MORE

Selected references for the ideas presented in this book are given below. Some references are specific to the case study examples, others relate to general principles.

Chapter 1: The rationale for project evaluation

Sustainability: World Commission on Environment and Development (1987) *Our Common Future* (Brundtland Commission) (Oxford University Press, Oxford).

Economic model of sustainability: D Pearce, A Markandya & E P Barbier (1989) *Blueprint for a Green Economy* (Earthscan, London).

Chapter 2: Evaluating long-term projects

Historical review of explanations for time preference: M N Tothbard 'Time Preference', in J Eatwell, M Milgate & P Newman (eds) (1990) *The New Palgrave: Capital Theory* (Macmillan, London).

Recent analysis of time preference: Oxera Consulting Ltd (2002) *Social Time Preference Rate for Use in Long-Term Discounting* (UK Office of the Deputy Prime Minister, Department for Transport, and Department of the Environment, Food and Rural Affairs).

Discounting and present value: R A Brealey & S C Myers (2003) *Principles of Corporate Finance* (7th edn) (McGraw Hill / Irwin, New York), chapters 2, 3.

The standard approach to whole-life costing: R Flanagan, G Norman, J Meadows & G Robinson (1989) *Life Cycle Costing: theory and practice* (BSP Professional Books, Oxford).

More recent presentations of WLC: C Pasquire & L Swaffield 'Life-cycle/whole-life costing', in J Kelly, R Morledge & S Wilkinson (eds) (2002) *Best Value in Construction* (Blackwell, Oxford); R Flanagan & C Jewell (2004) *Whole-life Appraisal for Construction* (Blackwell, Oxford); Ivor H Seeley (1996) *Building Economics* (4th edn) (Macmillan, London).

Chapter 3: Evaluating risky projects

Current understanding of how people make economic decisions under uncertainty: C F Camerer, G Loewenstein & M Rabin (eds) (2004) *Advances in Behavioral Economics* (Princeton University Press, Princeton, NJ).

The impact of risk in financial decision-making: R A Brealey & S C Myers (2003) *Principles of Corporate Finance* (7th edn) (McGraw Hill / Irwin, New York), chapters 7-9.

Many examples of the unpredictable life-histories of buildings are given in: S Brand (1997) *How Buildings Learn: what happens after they're built* (Phoenix Illustrated, London).

Chapter 4: Data for whole-life costing

Methods of investment appraisal developed in business and finance since the 1950s: R A Brealey & S C Myers (2003) *Principles of Corporate Finance* (7th edn) (McGraw Hill / Irwin, New York), chapters 10-12.

Historical survey of the risk premium in American markets: R D Arnott & P L Bernstein (2002) 'What risk premium is "normal"?' *Financial Analysts Journal*, March/April.

Chapter 5: Lifecycle options

Historical survey of risk analysis and risk management, including the development of options theory: P L Bernstein (1996) *Against the Gods* (Wiley, New York).

Financial options and real options: R A Brealey & S C Myers (2003) *Principles of Corporate Finance* (7th edn) (McGraw Hill / Irwin, New York), chapters 20-22.

There are many books on real options, including:
M Amram & N Kulatilaka (1999) *Real Options: managing strategic investment in an uncertain world* (Harvard Business School Press, Boston, Mass.) – an introductory presentation;
T Copeland & V Antikarov (2001) *Real Options: a practitioner's guide* (Texere, New York) – more technical detail;
A K Dixit & R S Pindyk (1994) *Investment under Uncertainty* (Princeton University Press, Princeton, NJ) – a mathematical approach.
Also: S Howell, A Stark, D Newton, D Paxson, M Cavus & K Patel (2001) *Real Options: introduction for executives* (Financial Times/Prentice Hall, London).

Chapter 6: The option to develop

A classic account of commercial property development: O Marriott (1967) *The Property Boom* (Pan Books, London).

Technical discussion of the option to develop: S Titman (1985) 'Urban land prices under uncertainty' *American Economic Review* vol.175, no.3 (June), pp.505-514.

Chapter 7: The option to expand

Cantilevered extensions to high-rise flats in Hong Kong: J Wojtowicz (1984) *Illegal Facades* (Privately published, Hong Kong).

Leasing options: K Patel, D A Paxson & Sing Tien Foo (2000) *Real Urban Development Options at Canary Wharf* (Manchester Business School, Manchester).

Chapter 8: The option to switch use

Analysis of office and retail rents: G Matysiak & S Tsolacos (2001) *Identifying Short Term Real Estate Rental Performance: leading indicators* (Real Estate Finance and Investment Research Paper 2001.07, City University Business School, London); P D Childs, T J Riddiough & A Triantis (1995) *Mixed Uses and the Redevelopment Option* (Working Paper 56, Massachusetts Institute of Technology, Cambridge, Mass).

Designing for change in use: J Weeks (1969) 'Multi-strategy buildings' *Architectural Design* (October) pp.536-540.

Chapter 9: The option to reconfigure

The Wessex Water headquarters by Bennetts Associates: *Architecture Today* no.113 (November 2000) pp.60-76.

Risk management using credit derivatives: David Rule (2001) 'The credit derivatives market: its development and possible implications for financial stability' *Financial Stability Review* (Bank of England), June 2001.

Hong Kong emergency housing: E G Pryor (1983) *Housing in Hong Kong* (2nd edn) (Oxford University Press, Hong Kong).

Follow-up study of flexible housing schemes in the UK: R Pike & C Powell (2004) 'Housing flexibility revisited' *Made* (Welsh School of Architecture), issue 1, pp.64-71. This article has a bibliography of proposals for flexible housing.

Recent proposal for a kit of parts approach to reconfiguration: John E Fernandez (2003) 'Design for change: Part 1 – diversified lifetimes' *Architectural Research Quarterly* vol.7, no.2, pp.169-182.

Chapter 10: The option to refurbish

Market uncertainty and the specification for housing refurbishment: Arto Saari (1998) *The Viability of Rehabilitated Apartment Buildings* (Acta Polytechnica Scandinavica, Civil Engineering and Building Construction Series, no.111).

Design stage options for future upgrading: D J Fisk (1976) *Energy Conservation: energy costs and option value* Paper CP 57/76 (Building Research Establishment, Watford).

Chapter 11: Options on new technology

PV technology: Max Fordham & Partners (1999) *Photovoltaics in Buildings: a design guide* (Department of Trade & Industry, London).

Proposed legislation for 'solar ready' roofs: *Building*, 19 Sept 2003, p.14.

Chapter 12: Very long-term projects

Nineteenth century sources: Henry Fawcett (1876) *Manual of Political Economy* (4th edn) (Macmillan, London), republished 1995 by Thoemmes Press, Bristol; Rivington (1899) *Notes on Building Construction* (Longmans Green, London), republished 2004 by Donhead, Shaftesbury, Dorset.

The Albert Memorial and its refurbishment: C Brooks (2000) *The Albert Memorial* (Yale University Press, New Haven, CT).

Time-varying discounting: Oxera Consulting Ltd (2002) *Social Time Preference Rate for Use in Long-Term Discounting* (UK Office of the Deputy Prime Minister, Department for Transport, and Department of the Environment, Food and Rural Affairs); C F Camerer, G Loewenstein & M Rabin (eds) (2004) *Advances in Behavioral Economics* (Princeton University Press, Princeton, NJ); HM Treasury (2003) *The Green Book: Appraisal and Evaluation in Central Government* (The Stationery Office, London).

Chapter 13: Lifecycle budgeting under uncertainty

Data on component life: HAPM Publications Ltd (1992) *HAPM Component Life Manual* (Spon, London); Building Performance Group (2000) *BPG Building Fabric Component Life Manual* (Spon, London); N Harvey (2001) *Life Expectancy of Building Components: surveyor's experiences of buildings in use* (Building Cost Information Service and Royal Institution of Chartered Surveyors, London).

Method for estimating component life using weightings: K Bourke & H Davies (1997) *Factors Affecting Service Life Predictions in Buildings* Paper BR 320 (Building Research Establishment, Watford).

Statistical modelling of uncertain service life: P D T O'Connor (1991) *Practical Reliability Engineering* (3rd edn) (Wiley, Chichester).

INDEX

References to principle boxes are in bold type.